Multi-physics Optimization

Reliability of Multiphysical Systems Set

coordinated by
Abdelkhalak El Hami

Volume 20

Multi-physics Optimization

Mechanics, Fluid Interaction Structure, Shaping, Stochastic Finite Elements, Random Vibrations, Fatigue

Abdelkhalak El Hami
Bouchaïb Radi

iSTE

WILEY

First published 2025 in Great Britain and the United States by ISTE Ltd and John Wiley & Sons, Inc.

ISTE Ltd
27-37 St George's Road
London SW19 4EU
UK

www.iste.co.uk

John Wiley & Sons, Inc.
111 River Street
Hoboken, NJ 07030
USA

www.wiley.com

Library of Congress Control Number: 2025938441

British Library Cataloguing-in-Publication Data
A CIP record for this book is available from the British Library
ISBN 978-1-83669-031-3

The manufacturer's authorized representative according to the EU General Product Safety Regulation is Wiley-VCH GmbH, Boschstr. 12, 69469 Weinheim, Germany, e-mail: Product_Safety@wiley.com.

Contents

Preface

Optimization is a field of mathematics in which the focus of interest is on maximizing (or minimizing) a given objective, such as energy or an economic value. Problems in dynamics of mechanical structures, for example, are of great importance; they lead in particular to building mechanical assemblies that are increasingly more flexible and subject to excitations that fluctuate more quickly over time.

This book illustrates, in great detail, the state of the art of the multidisciplinary science of multi-physics optimization and structure reliability. In a context of perpetual research to improve industrial competitiveness, the evolution of methods and tools for designing and optimizing products becomes a strategic need when cost reduction is imperative. To increase the competitiveness of their mechatronic devices, equipment manufacturers in civil engineering, automotive and aerospace industries must innovate in both design and assembly processes to reduce product development times. In addition, these innovating products must naturally combine excellent functional and operational performance, including reliability optimization, in order to fully meet the expectations of the global market. In the field of aeronautics, the needs mainly relate to the forecasting and control of the costs induced by failures appearing during the initial period of operation, during the warranty period and during the normal operation of the aircraft, bearing in mind that, in the future, new contracts for the sale of aeronautical equipment will be more oriented toward sales, based on the time of operation.

Although the aeronautics sector features relatively low production volumes compared to the automotive sector (in terms of the number of units per type of product), the financial challenges are significant and as such, aircraft manufacturers oversize mechatronic components to deal with unknown problems.

This book proposes new optimization methods coupled with reliability methods for the analysis of the life cycle of a system. The V-cycle allows for the inserting of development phases into the period of time needed for development and the levels of integration complexity. Each chapter begins with reminders of essential results, which should not prevent the reader from consulting the references at the end of the book.

Chapter 1 presents the different methods in mechanics for structural dynamics analysis and optimization. Then, the notion of uncertainty is taken into account by presenting the different stochastic finite element and reliability optimization methods.

Chapter 2 presents a methodology coupling the techniques of reliability optimization in design and modal synthesis methods. First, we present the state of the art in modal synthesis methods. We then propose an algorithm by integrating modal synthesis methods into the reliability optimization process. Finally, we evaluate this algorithm with different applications by showing the efficiency and robustness of the proposed method.

Chapter 3 covers the presentation of the general equations of the dynamics of mechanical structures by considering their application to stochastic modal synthesis methods. The first section covers the equations of motion for a mechanical structure, the notations for the calculation of dynamic responses, the computation of eigenmodes and the frequency response functions, as problems. On the other hand, direct resolution methods integrate equations of motion in order to be able to work with nonlinear structures or if the frequency content of the excitation covers a large number of structural modes. Later in the chapter, the method of substructuring is recalled, which consists of treating a structure as an assembly of interconnected substructures.

The objective of Chapter 4 is to determine the best methodology to follow for the reliability optimization of structures subjected to random vibrations, taking into account fatigue damage.

Chapter 5 presents optimization and shaping by hydroforming. This process involves several complex phenomena and exhibits several types of nonlinearities (material, geometric, boundary conditions, etc.). The development of such an operation requires a lot of testing in order to determine the optimum loading pathways in an accurate manner and to obtain a part with no defect.

Chapter 6 presents the problems of reliability optimization in the design of structures in vibration and in interaction with a flowing fluid in order to detect the critical frequency bands that can lead to the damage (or destruction) of the structure to be optimized.

Chapter 7 is dedicated to the presentation of the generalized polynomial chaos method in the process of reliable dental implant optimization.

Acknowledgments

We would like to thank all of the people who have directly or indirectly contributed to the development of this book, in particular the engineering students and doctoral students of the INSA of Rouen Normandy, who have been under our supervision for the past few years.

Abdelkhalak EL HAMI
Bouchaïb RADI
June 2025

1

Introduction to Optimization in Mechanics

1.1. Introduction

The purpose of this chapter is to present a nonexhaustive state of the art of all of the fields that will be directly or indirectly covered in this book.

The first section of this chapter is dedicated to the general study of structural dynamics. This study will make it possible to define the notations essential to the calculation of dynamic responses, to the calculation of frequencies, eigenmodes and response functions. All of these aspects will be addressed in detail and through practical applications.

One of two conventional strategies can be used to solve the system of dynamical equilibrium equations of a structure [IMB 95]. The most common solution strategy in dynamic range is modal superposition, which is suitable for linear structures in which only the first modes are excited. On the other hand, direct solving methods use the integration of equations of motion, in order to deal with nonlinear structures. The latter can also be applied when the frequency content of the excitation covers a large number of modes of the mechanical structure under study.

The second section of this chapter is a presentation of a bibliographic study of structural optimization. The goal is to obtain the appropriate shapes for a part by minimizing a given criterion. In every area of structural mechanics, the impact of the proper design of a part is very important for its strength, service life and application use. This challenge is a daily one in high-tech sectors. The development of the art of engineering requires considerable effort to constantly improve structural design techniques. Optimization is essential in increasing performance and reducing the weight of aerospace and automotive vehicles, thus leading to substantial energy savings.

The last section of this chapter is dedicated to the description of the different tools for structural analysis with uncertain parameters. The taking into account of parameter uncertainty is a particularly critical problem in vibrational mechanics. However, the consideration of this effect can respond to different types of needs, where two main categories can be distinguished: analysis and design. In general, the modeled parts or structures must satisfy given specifications, which require, for example, that safety, reliability or comfort standards be observed.

During deterministic design, engineers try to find the best possible design among all of the potential solutions they have to study. This choice is based on cost as well as on improving product quality. In this case, the designer's objectives to get an optimal design are devised without worrying about the accuracy of the mechanical characteristics of the materials, the geometry and loading (uncertainty effects). The resulting optimal design may therefore represent an inadequate level of reliability. The objective of the approach integrating reliability analysis into the optimization problem and called RBDO (for Reliability Based Design Optimization) is to design structures while establishing the best compromise between cost and the proper functioning of the device.

1.2. Problems of general dynamics

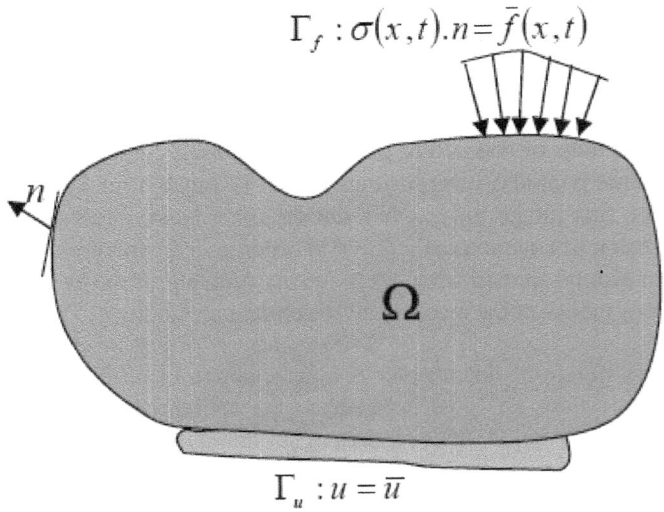

Figure 1.1. *System Ω subject to forces. For a color version of this figure, see www.iste.co.uk/elhami/multi-physics.zip*

The formulation of a problem of dynamics in small perturbations on a domain Ω of boundary $\Gamma = \Gamma_u \cup \Gamma_f$ (see Figure 1.1) and in a time interval [0, T] is:

$$Div_x \, \sigma(x,t) + g(x) = \rho \ddot{u}(x,t) \quad [1.1]$$

$$\varepsilon = \frac{1}{2}\left(\nabla u + \nabla^t u\right) \quad [1.2]$$

Initial conditions:

$$u(x,0) = u_0(x) \;\; x \in \Omega \quad [1.3]$$

$$\dot{u}(x,0) = \dot{u}_0(x) \;\; x \in \Omega \quad [1.4]$$

Boundary condition:

$$u(x,t) = \bar{u}(x,t) \;\; (x,t) \in \Gamma_u \times [0,T] \quad [1.5]$$

$$\sigma(x,t).n = \bar{f}(x,t) \;\; (x,t) \in \Gamma_f \times [0,T] \quad [1.6]$$

u is the displacement vector, σ and ε are respectively the stress and strain tensor. ρ is the density. Vectors g, $\bar{f}$ and $\bar{u}$ respectively represent the volume force, the external force and the imposed displacement. $\vec{n}$ is the normal vector to the surface.

In the case of an isotropic elastic domain, the law of behavior is written as:

$$\sigma_{ij} = \lambda \varepsilon_{kk} \delta_{ij} + 2\mu \varepsilon_{ij} \quad [1.7a]$$

λ and μ are the functions of Young's modulus and Poisson's ratio ν:

$$\lambda = \frac{E\nu}{(1+\nu).(1-2\nu)} \quad [1.7b]$$

$$\mu = \frac{E}{2(1+\nu)} \quad [1.8]$$

The dynamic problem presented above for this elastic example can be represented by the Navier equation as follows:

$$\mu\nabla^2 u_i(x,t)+(\lambda+\mu)\frac{\partial}{\partial x_i}(\nabla.u(x,t))=\rho\ddot{u}(x,t) \qquad [1.9]$$

where ∇^2 denotes the Laplacian operator: $\nabla^2=\frac{\partial^2}{\partial^2 x_1}+\frac{\partial^2}{\partial^2 x_2}+\frac{\partial^2}{\partial^2 x_3}$ and "∇" is the notation of the divergence operator: $\nabla\cdot u=\frac{\partial u_1}{\partial x_1}+\frac{\partial u_2}{\partial x_2}+\frac{\partial u_3}{\partial x_3}$.

1.2.1. *Finite element methods*

For structures with complex geometries, numerical methods such as the finite element method are used. In an elastodynamic problem, the displacements are generally expressed by a combination of vectors [GMÜ 97]:

$$u(x,t)=[B(x)]\{q(t)\} \qquad [1.10]$$

where $[B(x)]$ is the matrix of shape functions, and $\{q(t)\}$ is the vector of discrete real displacements, whose components are the discrete unknowns of the approximation.

In general, a structural mechanics problem, after discretization, is described by a system of second-order equations:

$$\begin{cases}[M]\{\ddot{q}(t)\}+[C]\{\dot{q}(t)\}+[K]\{q(t)\} = F(t)\\ \{q(0)\}=\{q_0\},\ \ \{\dot{q}(0)\}=\{\dot{q}_0\}\end{cases} \qquad [1.11]$$

where N is the number of degrees of freedom of the system; $M(N\times N)$ is the positive definite symmetric mass matrix, $C(N\times N)$ and $K(N\times N)$ are the viscous damping and stiffness non-negative definite symmetric matrices respectively. F is the vector of the applied forces.

Mathematically, equation [1.11] represents a system of second-order differential equations that can be solved either by a direct integration method or by the mode superposition method.

1.2.2. *Modal assumption method*

We apply the following modal transformation to system [1.11]:

$$\{q\} = [\Phi]\{p\} \qquad [1.12]$$

$\{p\}$ is the vector of generalized coordinates; $[\Phi]$ is the modal matrix verifying the orthogonality properties: $[\Phi]^T[M][\Phi] = I$ and $[\Phi]^T[K][\Phi] = \left[w^2\right]$ with: $\left[w^2\right] = diag\left[w_1^2\ w_2^2 \cdots\right]$, where w_i is the natural angular frequency, equation [1.12] reverts to:

$$\{\ddot{p}\} + [\Phi]^T[C][\Phi]\{\dot{p}\} + \left[w^2\right]\{p\} = \{P\} \qquad [1.13]$$

where $\{P\} = [\Phi]^T\{F\}$ is the modal force vector.

It can be assumed that the damping matrix is proportional to the mass and stiffness matrix. This assumption, known as the Rayleigh assumption, is quite commonly used in structural analysis. It therefore postulates that:

$$[C] = \alpha[M] + \beta[K] \qquad [1.14]$$

$$[\Phi]^T[C][\Phi] = \alpha[I] + \beta\left[w^2\right] \qquad [1.15]$$

which amounts to writing that:

$$\{\ddot{p}\} + \left(\alpha[I] + \beta\left[w^2\right]\right)\{\dot{p}\} + \left[w^2\right]\{p\} = \{P\} \qquad [1.16]$$

The decoupled system becomes:

$$\ddot{p}_i + 2\zeta_i\, w_i\, \dot{p}_i + w_i^2\, p_i = P_i \qquad [1.17]$$

$$2\zeta_i\, w_i = \alpha + \beta\, w_i^2\,, \quad i = 1, 2, \cdots\cdots, N \qquad [1.18]$$

The factor ζ_i is called the reduced damping coefficient for the i-th mode. α and β are the initially unknown values and are calculated from the reduced damping coefficient ζ_i.

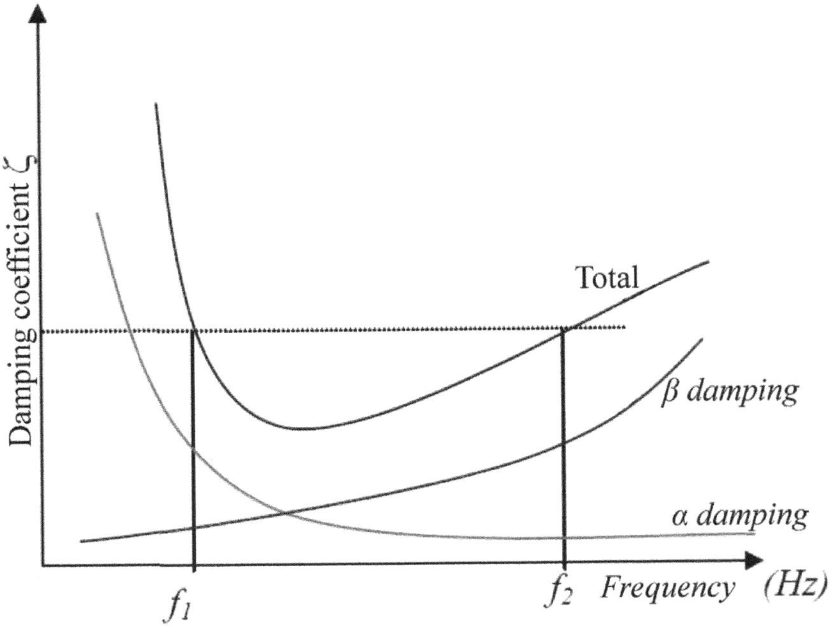

Figure 1.2. *Damping coefficient graph. For a color version of this figure, see www.iste.co.uk/elhami/multi-physics.zip*

In Figure 1.2, we graphically present the modal damping coefficient ζ; note that the sum of the two functions is almost constant at damping over the frequency range of interest. Thereby, given the modal damping (ζ) and a frequency interval (f_1 and f_2), the two equations can be simultaneously solved to determine α and β.

$$\frac{\alpha}{4\pi f_1} + \beta\pi f_1 = \zeta \qquad [1.19]$$

and

$$\frac{\alpha}{4\pi f_2} + \beta\pi f_2 = \zeta \qquad [1.20]$$

1.2.3. *Direct integration*

There are many methods to integrate differential equations. The general procedure consists of discretizing time and formulating what happens at time "t+Δt" as a function of what happens at time "t" based on Taylor expansions; in this part, we present the Newmark method and the Wilson method.

1.2.3.1. *Newmark method*

Newmark proposed a method where the velocity and displacements at t+Δt are estimated with respect to $\{\ddot{q}_t\},\{\dot{q}_t\},\{q\}$ and to accelerations $\{\ddot{q}_{t+\Delta t}\}$. In addition, the displacement and velocity are developed in the Taylor series using the two independent parameters β and γ as well as the time step [KLE 92]:

$$\{q_{n+1}\}=\{q_n\}+\{q_n\}\Delta t+\left\{\left(\frac{1}{2}+\beta\right)\ddot{q}_n+\beta\ddot{q}_{n+1}\right\}\Delta t^2 \quad [1.21]$$

$$\{\dot{q}_{n+1}\}=\{\dot{q}_n\}+\{(1-\gamma)\ddot{q}_n+\gamma\ddot{q}_n\}\Delta t \quad [1.22]$$

where $\{q_n\}$,$\{\dot{q}_n\}$ and $\{\ddot{q}_n\}$ are the approximations of $\{q(t_n)\}$, $\{\dot{q}(t_n)\}$ and $\{\ddot{q}(t_n)\}$ respectively, and $t_{n+1} = t_n + \Delta t$, with Δt being the time steps, the two independent parameters β and γ ensure the accuracy and stability of the solution. When $\gamma \geq 1/2$ $\beta \geq (\gamma + 0.5)/4$.

By carrying these equations forward into the equation of motion, the following matrix relation is obtained:

$$\left(M+\Delta t\,\gamma C+\Delta t^2\,\beta K\right)\ddot{q}_{n+1}=C\,\dot{q}_{n+1}+K\,q_{n+1}-F_{n+1} \quad [1.23]$$

with $\dot{q}_{n+1}=\dot{q}_n+(1-\gamma)\Delta t\,\ddot{q}_n$ and

$$q_{n+1}=q_n+\Delta t\,\dot{q}_n+\Delta t^2\left(\frac{1}{2}-\beta\right)\ddot{q}_n \quad [1.24]$$

The acceleration at time t = 0 is provided by the equilibrium conditions and the initial conditions on $\{q\}$ and $\{\dot{q}\}$. The solution of equation [1.23] requires that a linear system be solved at each time step.

1.2.3.2. *The Wilson-θ method*

This is an extension of a method in which the acceleration is assumed to linearly vary over the interval $\left[n\Delta t,(n+1)\Delta t\right]$; Wilson [KLE 92] assumes that this linear variation occurs over the interval $\left[n\Delta t,(n+1)\Delta t\right]$. θ is a parameter whose value recommended by Wilson is 1.4.

If τ denotes time in the interval $[0,\theta\Delta t]$, then the acceleration in the interval $[t,t+\theta\Delta t]$ is written [KLE 92] as:

$$\ddot{q}_{t+\tau}=\ddot{q}_t+\frac{\tau}{\theta\Delta t}\left(\ddot{q}_{t+\theta\Delta t}-\ddot{q}_t\right) \qquad [1.25]$$

Speed and displacement are obtained by successive integrations

$$\dot{q}_{t+\tau}=\dot{q}_t+\tau\ddot{q}_t+\frac{t^2}{2\theta\Delta t}\left(\ddot{q}_{t+\theta\Delta t}-\ddot{q}_t\right) \qquad [1.26]$$

$$q_{t+\tau}=q_t+\dot{q}\tau+\frac{1}{2}\ddot{u}_t\tau^2\,\frac{\tau^3}{6\theta\Delta t}\ddot{q}_{t+\theta\Delta t}-\ddot{q}_t \qquad [1.27]$$

These basic equations were generalized by Hughes [TRO 87]. They are given for time $t=\theta\Delta t$ and with the notation $q(n\Delta t)=q_n$:

$$M\,\ddot{q}_{n+\theta}+C\,\dot{q}_{n+\theta}+K\,q_{n+\theta}=F_{n+\theta} \qquad [1.28]$$

$$\ddot{q}_{n+\theta}=(1-\theta)\ddot{q}_n+\theta\ddot{q}_{n+1} \qquad [1.29]$$

$$\dot{q}_{n+\theta}=\dot{q}_n+\theta\Delta t\left[(1-\gamma)\ddot{q}_n+2\beta\ddot{q}_{n+\theta}\right] \qquad [1.30]$$

$$q_{n+\theta}=q_n+\theta\Delta t\,\dot{q}_n+\frac{(\theta\Delta t)^2}{2}\left[(1-2\beta)\ddot{q}_n+2\beta\ddot{q}_{n+\theta}\right] \qquad [1.31]$$

$$F_{n+\theta}=(1-\theta)F_n+\theta F_{n+1} \qquad [1.32]$$

These equations are equal to the previous ones for the values $\beta = 1/6$ and $\gamma = 1/2$.

1.3. Structural optimization

Structural optimization is not a recent concern. If we search the archives, we find a calculation made by Galileo that gives the law of thickness of a free embedded beam assuming a constant stress distribution [TRO 87]. The most remarkable thing is that the result of this calculation is entirely confirmed by current theories. Obviously since Galileo, thousands of publications have been produced, and if we only consider the modern era of optimization, it is found that since 1960, Schmit has introduced the idea of coupling structural analysis by finite elements and nonlinear mathematical programming to search for the different automated optimal designs. As a matter of fact, he proved that structural design could be formulated by a mathematical programming problem [SCH 84].

The determination of the proper shape of structural components is a major engineering issue. In every area of structural mechanics, the impact of the proper design of a part is very important for its strength, service life and operation use. This can be seen as a daily challenge in the cutting-edge sectors of space research, aeronautics, automotive, naval competition, fine mechanics, precision mechanics or civil engineering structures, etc. The development of the art of engineering requires considerable effort to constantly improve structural design techniques. Optimization plays a key role in increasing performance and reducing the weight of aerospace and automotive vehicles, resulting in substantial energy savings.

The constant development of computer-aided design techniques and optimization strategies has to be viewed in this context. Structural optimization has been the focus of great interest for more than 20 years. Still not applied enough to traditional techniques of engineering departments, it is gradually being integrated into them as its reliability increases. Starting from the simplest problems, the field of application of structural optimization is now expanding to new and more interesting challenges.

To illustrate the evolution of structural optimization techniques, we can arbitrarily divide structure optimization into three main families: design optimization, shape optimization and topology optimization. Historically, each has been approached in increasing order of difficulty and generality [DUY 96].

1.3.1. *Design optimization*

This enables the improvement of a structural model while following the available resources (called constraints or limitations). The automated sizing of structures is the special case in which only the cross-section or cross-sectional thickness of the components of a structure whose shape and topology are preset can be modified [AZI 02], [BOU 01a]. No modification of the geometry model is possible.

1.3.2. *Shape optimization*

Shape optimization allows for shape changes that are compatible with a predetermined topology. In its conventional form, optimization modifies the parametric representation of the boundaries of the domain [ZHA 92, AFO 02]. By moving the boundaries of the domains, a better solution can be looked for among all of the structures obtained by homeomorphic transformation of the original structure (see Figure 1.3). In this case, it is clear that a change in the transverse dimensions as well as a modification of the configuration of the structure are possible, but altering the connectivity or the nature of the structural members is certainly not admissible.

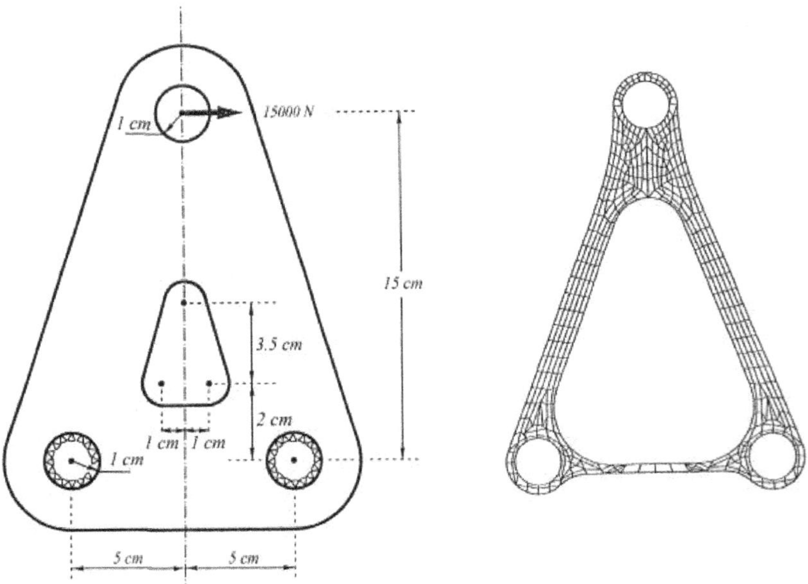

Figure 1.3. *Shape optimization: the initial model and the solution of the problem after five iterations [ZHA 92]*

1.3.3. *Topological optimization*

Topological optimization makes more fundamental modifications possible to the nature of the structure. This time, the geometry of the part is considered without any presumptions about the connectivity of the domains or structural members present in the solution. Optimizing the topology naturally leads to determining the optimal shape or transverse dimensions of the structure in a certain way, so that some authors [ROZ 93] also refer to it as generalized shape optimization. The final structure must satisfy the stresses defined beforehand by the user (usually related to the restriction of the maximal von Mises stress) [KIM 00].

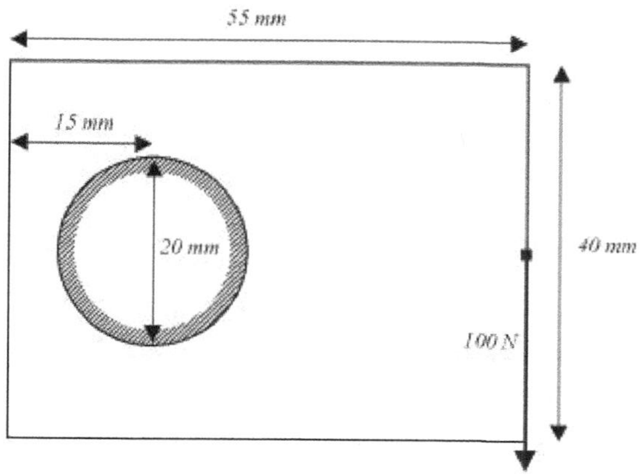

Figure 1.4. *Definition of the lattice problem by Michell [REY 99]*

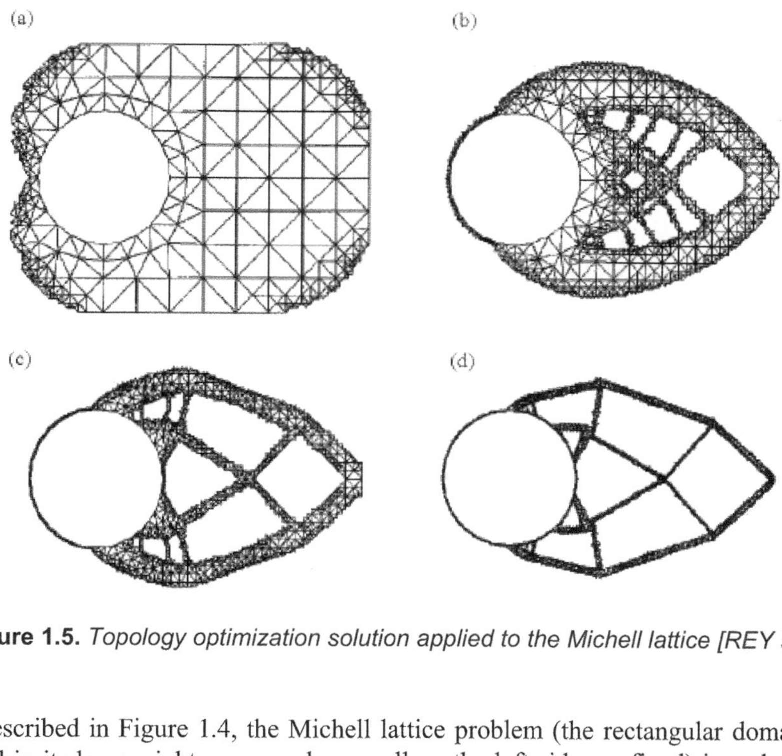

Figure 1.5. *Topology optimization solution applied to the Michell lattice [REY 99]*

Described in Figure 1.4, the Michell lattice problem (the rectangular domain is loaded in its lower-right corner, whereas all on the left side are fixed) is a classical reference in topological optimization. [DEB 99] solved the problem by an adaptive change technique, which consists of the following: once the initial finite element problem is established, the method begins with a mesh refinement by the subdivision of elements of regions where the stress (von Mises) is minimal. Then, the subdivided elements have a minimal stress and will be eliminated, etc., until convergence.

The solution of the structures obtained after 6, 42, 75 and 120 iterations of this process respectively is represented in Figure 1.4. At the 120th iteration, only 8.8% of the entire (initial) sector remains.

Topological optimization can also be performed for lattices. For example, [DEB 99] described a method to find the optimal cross-sections and topology of a lattice for 2D and 3D using a genetic algorithm. The aim was to reduce mass to a minimum, under constraints of forces and displacements that have predefined values. Figure 1.6 shows an example of a 3D lattice where the genetic algorithm converged to a nine-element lattice (from a 39-element configuration).

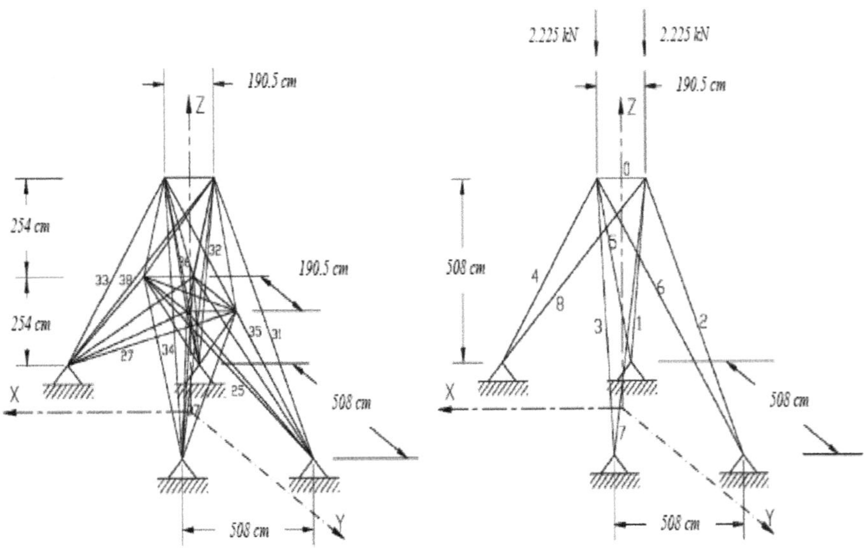

Figure 1.6. *Example of topology optimization [DEB 99]*

Recent developments in the coupling of shape and topological optimization methods by the finite element method can be found in [MAC 03].

1.3.4. *Definitions and formulation of an optimization problem*

An optimization problem is usually formulated as a minimization problem and written as:

$$\begin{cases} \min\limits_{x} f(x) \\ \textit{such that} \\ g_i(x) \le 0, i = 1 \cdots\cdots, m, \\ h_j(x) = 0, j = 1 \cdots\cdots, p, \\ x \in S \subset \Re^n, \end{cases} \qquad [1.33]$$

where f is the scalar function to be minimized, called *the cost function* or *objective function*, x is *the vector of the optimization variables*, g_i are the inequality *constraints* and h_j are the equality constraints, and S is *the space of variables* (also called the search space). S indicates what type of variables are considered: real, integer, mixed (real and integer in the same problem), discrete, bounded continuous, etc.

A point x_A is called an *admissible point* if $x_A \in S$ and if the optimization constraints are satisfied: $g_i(x_A) \le 0, i = 1 \cdots\cdots, m$ and $h_j(x_A) = 0, j = 1 \cdots\cdots, p$. The solution of (I.40) is the set of optima $\{x^*\}$.

x^* is a global minimum of f if and only if $f(x^*) \le f(x) \ \forall \ x \in S$.

x^* is a local minimum of f if and only if $f(x^*) \le f(x) \ \forall \ x \in S / \|x - x^*\| \le \varepsilon$, $\varepsilon > 0$.

Figure 1.7 shows an example of a single-variable function, with local minima and a global minimum. Among the local minima, the one with the smallest value of f is the global minimum.

A multimodal function exhibits several (local) minima, and a unimodal function has only one minimum, the global minimum. Figure 1.8 shows a multimodal function with two variables.

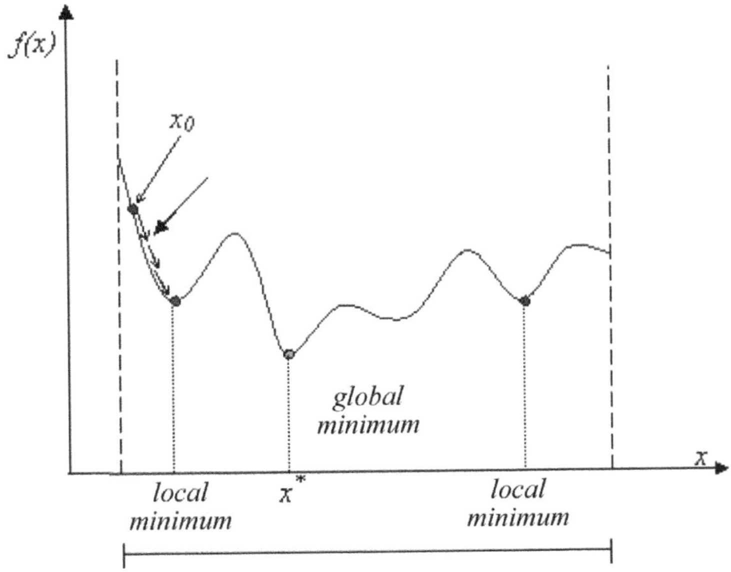

Figure 1.7. *Local minima and global minimum of a function with one variable. For a color version of this figure, see www.iste.co.uk/elhami/multi-physics.zip*

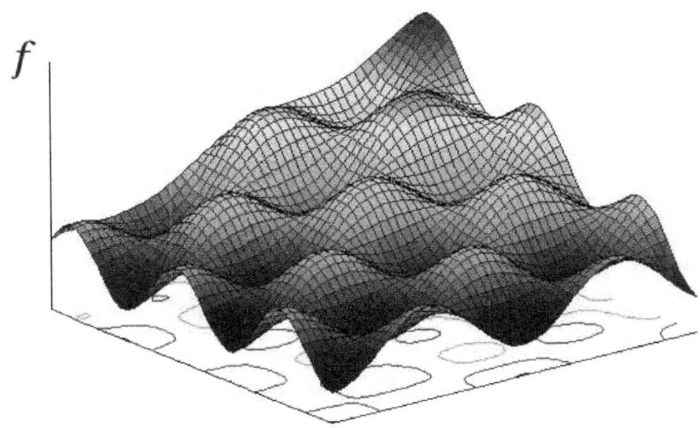

Figure 1.8. *A multimodal function with two variables. For a color version of this figure, see www.iste.co.uk/elhami/multi-physics.zip*

A local (or algorithm or search) method is that which converges to a local minimum. Local research usually starts from an initial point χ_0 with an initial step Δ_0. These parameters will condition the descent of one of the valleys of the function (see diagrams in Figure 1.7). There are many existing local methods. The oldest and most widely used are the methods where the direction of descent is deduced from the derivatives of the function (steepest slope method, Newton's method, conjugate guardian method, quasi-Newtonian methods [MIN 86]).

The aim of global methods is to achieve one or more global optima. Typically, the machine cost of an optimization is conditioned by the number of evaluations of the objective function [MOH 04]. If, for example, a mechanical system modeled by finite elements is to be optimized, the computation time will mainly be the sum of the time of the simulations used by the optimizer.

1.4. Structures with uncertain parameters

Accounting for uncertainties when studying a structure must take into account the two types of sources of uncertainty: those concerning excitation and those concerning structure. Dessombz [DES 00] classified uncertainties into four main categories based on Prasthofer's work: random parameters, poorly known parameters, variable parameters and model uncertainties:

a) *Random parameters*. These are, for example, the dimensions of a part whose tolerance is known, or Young's modulus or the density of a material such as steel.

b) *Poorly known parameters*. The case of boundary conditions is a typical problem. An embedment corresponds to a stiffness with a very high value, but of which only an order of magnitude is known at best. Similarly, the different types of assembly, such as welding, gluing, etc., are difficult to model, and the deterministic values used to represent these phenomena appear to be largely insufficient.

c) *Variable parameters*. We can mention the parameters that can be variable over time, and that cannot be controlled or not very well (typically material degradation or aging). We can also acknowledge parameters whose value can be known at a given time. In this case, it may be an excitation force (a train crossing over a bridge), a quantity of fuel in a tank or even special settings.

d) *Model uncertainties*. The chosen laws of behavior, for example, are the ones that poorly or incompletely represent physical phenomena, the errors due to the choice of the finite element mesh, its finesse and the elements chosen. These uncertainties are in general difficult to assess.

To represent parametric uncertainty, different approaches, probabilistic theory [IBR 87] [SAR 11], fuzzy sets [CAM 00] and convex models [BEN 90] have been proposed in the literature, to which the reader should refer for more details concerning the main methods used in the treatment of uncertainties which are presented here, without going into details.

1.4.1. *Monte Carlo simulation*

For addressing uncertainty problems, whether experimental or numerical, the approach that was naturally used in the first place is based on the principle of sampling: Monte Carlo simulation, known and used intensively in several scientific fields, was introduced into structural mechanics by Schinozuka. The method is almost chosen as a reference and is both the simplest and the costliest.

The method is comprised of three stages. The first step is the most important; it consists of generating, by numerical simulation, a large set of realizations associated with the stochastic parameters of the physical problem. The next step is to solve the deterministic problem for each of the population elements, in order to obtain a population corresponding to the random response quantities of the system under study. In the last step, a large number of outputs is obtained, from which different static values of the interesting response variables can be computed.

The computing power required to perform all of the simulations severely restricts the applications of this method, which is why approximation techniques are preferred.

1.4.2. *Analytical method*

The different methods for analyzing uncertainty problems are in particular the perturbation method and the Neumann series decomposition. The principle of these methods is recalled in [IBR 87]. The use of Neumann series is based on finding the inverse of the operator of the problem by decomposing it into a Neumann series. This limits its application to certain types of differential equations, whereas the perturbation method does not have this limitation, since it corresponds to a (Taylor-type) decomposition of the output field. These methods are generally simple to implement, but only give good results for small perturbations and when the problems are linear.

1.4.3. *Stochastic finite element method*

Much of the current research deals with the problem of structures with uncertain parameters using stochastic finite element methods. These combine two techniques, namely "classical" finite element analysis, and statistical analysis. In general, we seek to determine the stochastic characteristics of random responses, based on knowledge of the hazards on: structural and geometric parameters, boundary conditions or loads in a system.

[SCH 01] outlined the various developments developed in this field, and proposed numerous references for each of the aspects of this problem. An explanation of the stochastic finite element principle described by [GHA 96] involves the use of a serial decomposition of stochastic processes, truncated to a certain order according to the desired precision. There are two levels of description of the stochastic aspect of the problem: the first consists of considering the characteristics of the structure as known stochastic fields. In this case, a Karhunen–Loeve decomposition [SOI 93] is generally used, which is analogous to the modal superposition used in structural calculation, since this decomposition has remarkable properties of orthogonality and convergence. The second level to be considered concerns the stochastic field formed by the solution at the nodes of the structure: in this case, the so-called polynomial chaos decomposition is used [ROG 91]. Therefore, stochastic solutions are projected on the basis of orthogonal polynomials whose variables are orthonormal Gaussians. The properties of this polynomial basis can be used to find either analytically or numerically the mean, standard deviation or distribution of the random solution. The main problem with this method lies in the choice of resolution algorithms and in storing the values. The possibility of implementing parallelism according to the characteristics of the matrices obtained are undoubtedly the essential aspects to be taken into account for the development of this method, which already gives satisfactory results, with a significant computation time.

A different approach to the stochastic formulation using the finite element method can be achieved by modifying the existing elements in order to allow the inclusion of defects; as such, [COM 01] present an axisymmetric element, based on a shell element. Nevertheless, it enables the analysis of structures including non-axisymmetric defects, such as a cylinder or cone radius, or a variable thickness on the circumference.

1.4.4. *Fuzzy logic method*

Fuzzy logic was founded by [ZAD 65]. Since its emergence, it has been the subject of several research projects. For [ZAD 94], fuzzy logic is fuzzy set theory (FST). FST is an extension of the theory of what is already known as "classical sets". The membership function of a classical set A is defined by:

$$\mu_A(x) = \begin{cases} 1 & \text{if } x \in A \\ 0 & \text{if } x \notin A \end{cases} \qquad [1.34]$$

This means that an element x is either in A $\left(\mu_A(x)=1\right)$ or not $\left(\mu_A(x)=0\right)$.

This definition makes it possible to make the link with classical probabilistic analysis, for which the definition does not differ much, except that A is in this case a point in R and not an interval of R. This difference, which is not easy to interpret from a practical point of view, leads to one that shows the limitations of probabilistic analysis: if an event A has a probability $p(A)$, then the probability of the opposite event of A is known and is equal to $p(\bar{A}) = 1 - p(A)$. This is not true in the theory of possibilities: if we denote by $P(A)$ the possibility of an event A, the relation that links this possibility to the possibility of the complementary event $\bar{A}$ is written as $P(A) + P(\bar{A}) \geq 1$, because the relation $\max\{P(A), P(\bar{A})\} = 1$ is always verified. This reflects the fact that if two events are considered to be opposite, one of them is always completely possible, and that if an event is considered possible, its constraint may very well be so as well. All this is extensively detailed in [DUB 88], in which all of the mathematical bases of the theory can be found, as well as many computational elements allowing for a first application of fuzzy arithmetic.

1.4.5. *Bibliography study of reliability methods*

Most of the approaches studied in structural reliability can be generalized into two technical categories of simulation and analytics. These techniques are used as a tool to assist the design engineer in taking into account all possible uncertainties during the design and construction phase in order to calculate the reliability index or probability of failure corresponding to one or more failure scenarios [MOH 04].

1.4.5.1. *Simulation technique*

Simulation techniques are a means to accurately evaluate the probability of failure. These techniques, also known as Monte Carlo techniques, involve different simulation methods. These include: direct Monte Carlo, importance sampling [ENE 93], conditional simulations, directional simulations, adaptive sampling and the Pavé method. The Monte Carlo simulation technique was originally proposed in the early 1940s for testing technology systems using the low-cost simulation technique. The application of the Monte Carlo simulation is performed for the calculation of the probability of failure.

In the Monte Carlo simulation [AYY 97], the failure probability is given by:

$$P_f = \frac{N}{N_f} \qquad [1.35]$$

where N_f is the number of failure events obtained and N is the total number of simulated events. The statistical precision of the failure probability is measured by its coefficient of covariance and is

$$\operatorname{cov}\left(P_f\right)=\sqrt{\frac{\left(1-P_f\right)P_f}{N}}\Big/P_f \qquad [1.36]$$

The equation indicates that a small failure probability implies a large number of simulation cycles to maintain an acceptable level of precision. Consequently, this results in increased computational costs. In addition, for a practical problem with several random variables, the Monte Carlo simulation becomes very expensive. To overcome this difficulty, several more effective alternative methods such as the method of importance sampling [AYY 97] have been developed. The fundamental principle for the method of importance sampling is to introduce random parameters with different distributions and that their averages are as close to the design point as that of the original distribution. As a result, simulation efficiency is increased since failure scenarios are obtained more frequently.

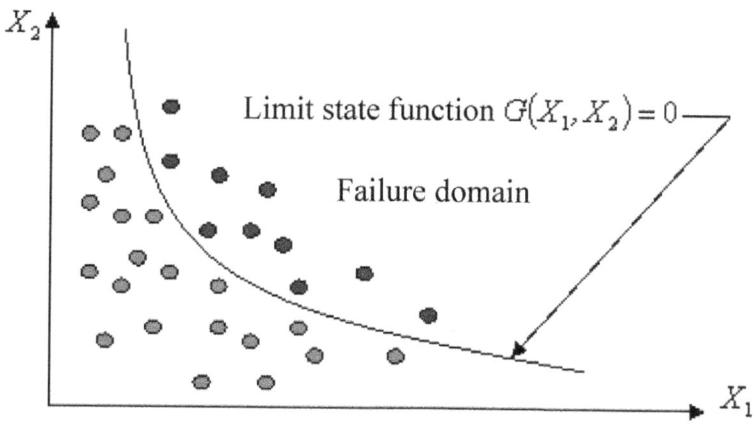

Figure 1.9. *Principle of the Monte Carlo method. For a color version of this figure, see www.iste.co.uk/elhami/multi-physics.zip*

[HAR 83] presented a general procedure based on the importance sampling technique for the calculation of the probability of failure; this procedure was applied to a fatigue problem. Therefrom, they showed the effectiveness of the importance sampling method compared to the Monte Carlo simulation.

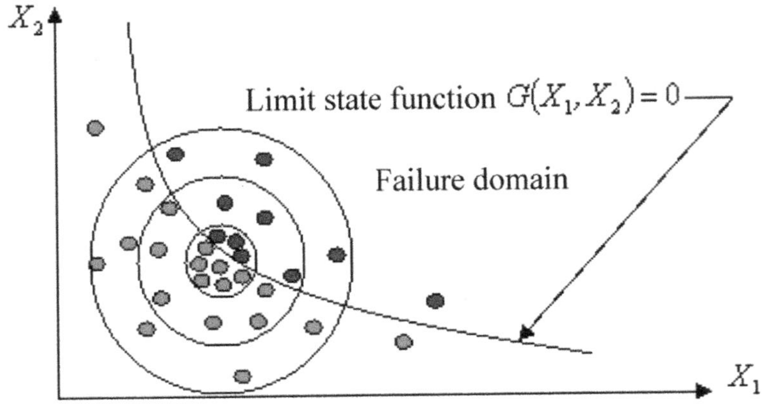

Figure 1.10. *Principle of importance sampling. For a color version of this figure, see www.iste.co.uk/elhami/multi-physics.zip*

1.4.5.2. *Analytical method*

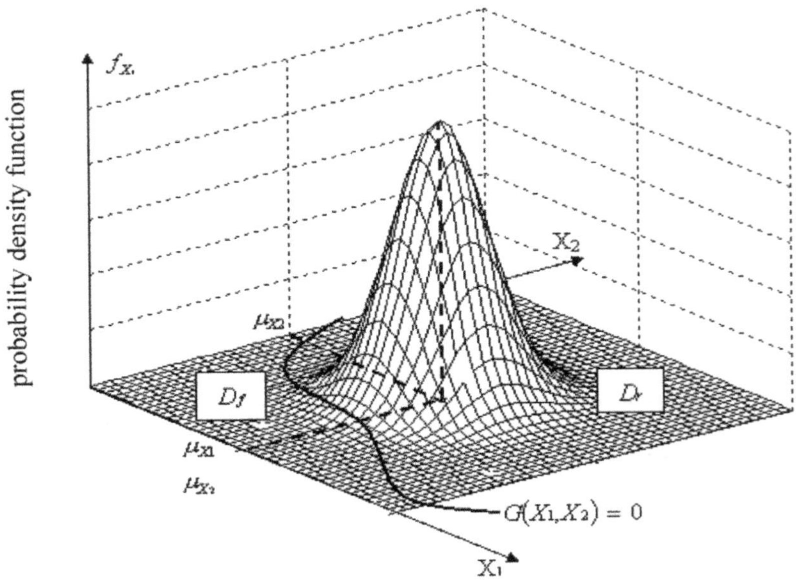

Figure 1.11. *Representation in the physical space of the probability density function for two variables*

Analytical techniques commonly used for structural reliability analysis include first-order and second-order reliability methods (FORM: First-Order Reliability Method; SORM: Second-Order Reliability Method). The fundamental concept behind these techniques consists of transforming random variables into statistically independent reduced Gaussian variables in a normalized space. Within this space, the FORM or SORM techniques are applied to approximate the limit state described by the failure criterion. The probability of failure is evaluated by a reliability index β such that $p_f \approx \Phi^{-1}(\beta)$, where Φ is the cumulative distribution function of the reduced centered normal distribution.

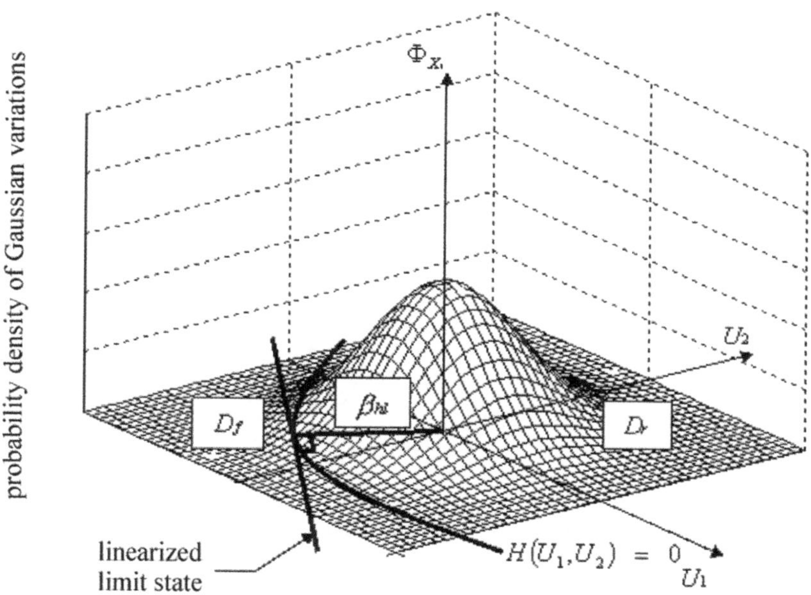

Figure 1.12. *Representation in the reduced space of the probability density function for two variables*

To evaluate the reliability index for the FORM method, [HAS 74] proposed an iterative algorithm in which the limit-state surface is approximated by a hyperplane tangent to the design point that corresponds to the most probable failure point (MMP: most probable failure point).

The reliability index is defined as the shortest distance between the origin of the axis system and the limit-state surface. [RAC 78] include information about the distribution. The [HAS 74] algorithms with that of [RAC 78] are called the HL-RF algorithm.

The FORM-based approach for calculating system reliability is due to [HOH 83]. They first reduced the system to a series of parallel systems, and then they gave a first-order solution to the multi-normal integral for the simple series or parallel systems. Finally, they determined the probability of failure.

[GOL 83] simplified the evaluation of system reliability by replacing the subsets with equivalent components within the FORM context.

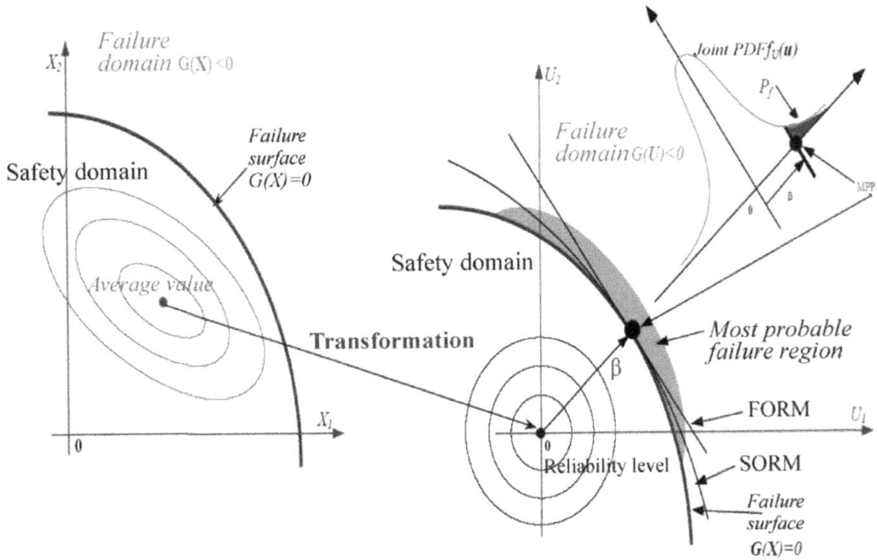

Figure 1.13. *Transformation of the probabilistic space into a normal space. For a color version of this figure, see www.iste.co.uk/elhami/multi-physics.zip*

[ENE 93] described a FORM-based method for predicting the system reliability index by modeling the serial system as a parallel system.

FORM is usually accurate for linear limit-state functions and not for strongly nonlinear limit states. To address this problem of limit-state nonlinearity, SORM was proposed to improve the evaluation of reliability by using a quadratic approximation of the limit-state surface.

Such an approximation was initially verified by [FIE 79]. Breitung [BRE 84] proposed an asymptotic method for predicting probabilities of failure for a large β by applying the quadratic approximation to MMP. He showed that the probability of failure can be expressed as:

$$P_f \approx \Phi(-\beta)\prod_{i=1}^{n-1}\left(1-\beta\frac{\varphi(-\beta)}{\Phi(-\beta)}k_i\right)^{1/2}, \beta \to \infty \qquad [1.37]$$

where k_i are the main curvatures of failure going through the MMP.

[TVE 90] offered an exact failure probability for a second-order approximation to MMP:

$$P_f = \frac{1}{2}-\frac{1}{\pi}\int_0^{\infty}\sin\left(\beta t+\frac{1}{2}\sum_{i=1}^{n-1}\tan^{-1}(-k_i t)\right)\frac{\exp\left(-\frac{1}{2}t^2\right)}{t\left[\prod_{i=1}^{n-1}\left(1+k_i^2 t^2\right)\right]^{1/4}}dt \qquad [1.38]$$

[DER 87] presented a paraboloid second-order approximation method around the most probable point. In another article, Der Kiureghian [DER 91] developed an efficient algorithm to determine the main curvature in an iterative way for the second-order reliability analysis. [KOY 94] proposed an alternative form to the second-order approach to reliability evaluation. It should also be mentioned that their algorithm can give accurate results for small or even negative values of β. [ZHA 98a] described a SORM process of which they first give a simple approximation; they then present a second-order reliability index to estimate the failure probability and finally evaluate the failure probability by a fast inverse Fourier transformation.

An improvement has been brought to Breitung's asymptotic formula by [HOH 87].

Although generally simulations or analytical techniques work very well for most problems, they could face computational problems when the limit-state function is not an explicit function of random variables but must be determined through complex structural analysis.

In this case, the response surface methodology (RSM) was used to obtain an analytical approximation of the implicit limit-state function, which is used by FORM/SORM or other approaches to estimate structural reliability. RSM is approved as a good approach for complex structural systems and particularly for those with complex limit states [SU 02].

In addition to the approaches mentioned, the artificial intelligence technique has emerged as a potential tool for the analysis of structural reliability. [SHA 97] proposed an approximation of the limit state using neural networks and they have also developed an active algorithm that allows the search for non-feasible regions.

1.4.6. *Bibliography study of reliability optimization methods*

[FRA 85] presented a sensitivity analysis technique that was later applied to design optimization in which the weight was taken as the objective function while the target reliability index was taken as constraint. [FRA 94] developed a vector optimization approach for structural design problems requiring multiple limit states and they are considered simultaneously. They also suggested a three-step reliability-based vector optimization strategy in the solution to the optimization problem.

[YAN 91] ran a reliability-based optimization system for an aircraft wing subjected to certain loads. FORM was used to predict the reliability index of various components, while Ditlevsen adopted a technique to obtain the reliability index of the system. A two-level optimization problem, in which weight was taken as an objective function and reliability index as a constraint, was presented by [YAN 90] for composite structures.

[ENE 93] suggested four different procedures to solve the problem of reliability optimization for serial and parallel systems. The first two approaches are based on sensitivity analysis and the last is based on sequential methods. In another article by the same authors [ENE 94], several aspects related to reliability optimization in structural engineering were discussed. A variety of reliability optimization problems were formulated; the FORM method was used to estimate the level of reliability of the system, and a two-level strategy was suggested to solve the reliability optimization problem. The choice of first-rate optimization algorithms at the same time as a sensitivity analysis increases the efficiency of the solution to the reliability-based optimization problem. They also proceeded with their discussion by examining several practical outcomes of reliability optimization including the use of finite element analysis. They also concluded the discussion with a description of a strategy for correcting and improving the model as well as evaluating the optimal outcome.

[ROY 01] presented a decoupling approach by which the optimization problem can be reformulated into a deterministic and semi-infinite problem (characterized by a finite number of design variables and an infinite number of constraints). This approach was then applied to the reliability optimization of serial structure systems with two optimization formulations. In the first formulation, cost is minimized under the constraint of reliability and mechanical stresses, whereas in the other, reliability is taken as an objective function under mechanical stress. The advantage of this approach lies in its flexibility as well as in the fact that any optimization algorithm and any reliability method can be independently adopted for the reliability-based optimization solution since computation, optimization and reliability are entirely decoupled.

Analytical techniques, mainly FORM/SORM, will always be a good choice for a reliability assessment in the reliability optimization problem, when the analytical model of the limit state function equivalent approximation is available. A one-level optimization strategy proposed by [KUS 97] based on FORM and using optimality criteria was used to solve both formulations of the reliability optimization problem. The first problem consists of minimizing the cost under reliability constraints. The second optimizes reliability under cost constraints. The algorithm of the approach was illustrated with three examples, and the optimization results were compared to those obtained by other methods available in the literature in terms of cost and stability. The disadvantage of this algorithm is that it is limited to a single limit state, so it is not applicable to reliability optimization problems.

[FEN 86] performed a structural optimization treating the probability of failure of the entire system as a constraint. An optimality criterion was used to solve the optimization problem.

For reliability optimization, two approaches, including the conventional reliability index and the target performance approach, were used to determine and satisfy the constraints in structural reliability. [LEE 02] conducted a comparative study, and they found that the target performance approach was computationally more efficient and robust than the conventional reliability index approach for the examples considered.

[WAN 96] established a set of methodologies that have applications in optimization problems based on system reliability. In order to accurately assess the failure probability for nonlinear limit-state functions, they proposed the HORM method (HORM: Higher-Order Reliability Method) to predict the probability of failure based on the reliability index obtained from an efficient safety index algorithm. The upper limit was then used to calculate the reliability of the system for which the commonalities are localized using a faster algorithm and using a high-order approximation. The approximation of two nonlinear points was developed to execute the reliability optimization system. The advantage of this method was demonstrated by a numerical test example.

[PAP 02] presented a methodology for reliability optimization where MCM with importance sampling was adopted to assess reliability while neural networks were used to perform the optimization.

The application of RSM by calculating reliability for reliability optimization is described in the article by [GAS 97], and the one by [SU 02].

An effective method was developed by [KHA 01, 02, 03] by merging the two physical and normal spaces. This approach is called the hybrid method. The hybrid approach consists of integrating the two sub-problems into a single problem that is simultaneously solved in the spaces of deterministic and random variables. Under this scheme, we integrate the reliability problem into the formulation of the optimization problem, in order to arrive at a single objective function and, thus, to reduce the overall computational cost. This last method was the objective of a detailed study during this thesis in order to develop methods adapted to the dynamic problem; among the most recent works in this field, we proposed a new method for the detection of critical areas in the vicinity of resonant frequencies as well as a new method called improved hybrid which allowed us to improve the results of the conventional hybrid method [MOH 05a]. These results will be explained in detail in the following chapters.

1.5. Conclusion

In this chapter, a literature review was presented in all of the areas that will be covered; in particular, reliability-based structural optimization. In the first section, the different analysis methods for structural dynamics were presented. In the second section, the notion of uncertainty was taken into account, different stochastic finite element methods were presented, as well as a historical study of reliability and optimization analysis.

2

Modal Synthesis and Reliability Optimization Methods

2.1. Introduction

Problems in structural dynamics are of great importance, in particular the tendency to build structures that are increasingly more flexible and subject to excitations that fluctuate increasingly rapidly over time. The inclusion of reliability optimization procedures for the design of complex dynamic systems (for large-scale problems) is becoming increasingly costly in terms of computational resources. However, a dynamic analysis of large industrial structures is usually based on model reduction techniques. The objective of this chapter is to present a methodology coupling modal synthesis techniques and reliable design optimization.

The chapter is divided into three paragraphs. We first present the state of the art of modal synthesis methods; and secondly, we propose an algorithm that allows us to introduce modal synthesis methods into the reliability optimization process. Finally, we evaluate this algorithm on different applications to show the efficiency and robustness of the proposed method.

2.2. State of the art of modal synthesis methods

2.2.1. *Introduction*

An analysis of the dynamic behavior of mechanical structures is essential for a reliable and optimal design. Such an analysis is usually based on finite element modeling. However, the cost becomes highly prohibitive for more complex or large models. Hence, the need to develop techniques to reduce the size of models while preserving their property. Different methods were developed, reducing the problems to a smaller size, by condensation methods, on the one hand, and splitting the analysis of large structures, on the other.

2.2.2. *Model reduction method for elastodynamic systems*

A structural problem of mechanics, after its discretization, is described by the following second-order system of equations:

$$M\,\ddot{x}(t) + C\,\dot{x}(t) + K\,x(t) = F(t) \qquad [2.1]$$

where N is the number of degrees of freedom (d.o.f.) of the system; $M(N{\times}N)$ is the positive definite symmetric mass matrix; $C(N \times N)$ and $K(N \times N)$ are the viscous damping and stiffness matrices that are nonnegative definite symmetric respectively. x and F respectively represent the displacement vector and the applied forces vector. The goal of model reduction methods is to find a new system of equations [CRA 68] with a number N_c much smaller than N $N_c << N$:

$$M_c\,\ddot{y}(t) + C_c\,\dot{y}(t) + K_c\,y(t) = F_c(t) \qquad [2.2]$$

where N_c is the number of d.o.f. of the system; $M_c(N_c \times N_c)$; $C_c(N_c \times N_c)$ and $K_c(N_c \times N_c)$ are the generalized matrices after condensation.

Modal synthesis methods aim to determine the vibration modes and response functions of large structures from the independent modal analysis of sub-parts. These methods are often the best and sometimes the only solution strategy for the dynamic analysis of complex structures.

The use of these methods is justified both by advantages in terms of their numerical efficiency and by taking into account the constraints of the organization of large projects. In addition, they make it possible to solve large dynamic problems with satisfactory accuracy.

The most commonly used techniques are condensation and modal synthesis methods [SAR 04], as shown in the following sections. We will also make reference to Craig Bampton's method and the junction d.o.f. reduction method.

2.2.3. *Fixed-interface method (Craig Bampton)*

In a harmonic analysis, the response $\{q\}e^{iwt}$ is related to the excitation $\{Q\}e^{iwt}$ by the equation:

$$\left[K^{(k)} - w^2 M^{(k)}\right] \{q\}^{(k)} = \{Q\}^{(k)} \qquad [2.3]$$

where *[K]* and *[M]* are the stiffness and mass matrices of the substructure *(k)* respectively.

The d.o.f. are decomposed into junction d.o.f. (slave(s)) and internal d.o.f. (master (m)).

The modal equation of motion of the complete system becomes:

$$\left(\begin{bmatrix} K_{ss} & K_{sm} \\ K_{ms} & K_{mm} \end{bmatrix} - w^2 \begin{bmatrix} M_{ss} & M_{sm} \\ M_{ms} & M_{mm} \end{bmatrix}\right)\begin{bmatrix} q_s \\ q_m \end{bmatrix} = \begin{bmatrix} 0 \\ Q_m \end{bmatrix} \qquad [2.4]$$

The vector $\{Q_m\}$ represents the inter-substructure connection forces.

In equation [2.4], the overall displacement $\{q\}^{(k)}\left(N^{(k)} \times 1\right)$ is split into master d.o.f. $\{q\}_m^{(k)}\left(m^{(k)} \times 1\right)$ and slave d.o.f. $\{q\}_s^{(k)}\left(s^{(k)} \times 1\right)$.

We introduce the Ritz approach which uses as basis vectors for each substructure the static modes of bonding and the truncated set of eigenmodes with fixed interfaces.

We therefore express the truncated displacements of the eigenmodes with fixed interfaces

$$q_s^{(k)} = \underbrace{\left(K_{ss}^{-1} K_{ms}\right)^{(k)} q_m^{(k)}}_{\left\{q_s^{(k)}\right\}^S} + \underbrace{\phi^{(k)}\eta}_{\left\{q_s^{(k)}\right\}^D} \qquad [2.5]$$

where $\left\{q_s^{(K)}\right\}^S$ is the static displacement of $\left\{q_s^{(k)}\right\}$ induced by $\left\{q_m^{(k)}\right\}$ and described by the static Guyan approximation [GUY 65].

$\left\{q_s^{(K)}\right\}^D$ is the dynamic motion of $\left\{q_s^{(k)}\right\}$ with η vector of modal coordinates, described by a subproblem by:

$$\left[K_{ss}^{(k)}\right]\varphi_v = \omega_v^2\left[M_{ss}^{(k)}\right]\varphi_v \qquad v = 1,2\cdots,s^{(k)} \qquad [2.6]$$

such as:

$$^T\Phi^{(k)}\,K_{ss}^{(k)}\,\Phi^{(k)} = \Sigma^{(k)} \text{ and } ^T\Phi^{(k)}M\,\Phi^{(k)} = I_s^{(k)} \qquad [2.7]$$

$\Phi^{(k)}\left(s_1^{(k)} \times s_1^{(k)}\right)$, $\Sigma^{(k)}\left(s_1^{(k)} \times s_1^{(k)}\right)$ are respectively the truncated modal base and the spectral matrix with $s_1^{(k)}$ computed mode $\left(s_1^{(k)} << s_1^{(k)}\right)$ of system [2.6]. Using relation [2.5], the vector *{q}* can be expressed by:

$$\{q\}^{(k)} = \begin{bmatrix} q_s \\ q_m \end{bmatrix} = \underbrace{\begin{bmatrix} -K_{ss}^{-1}K_{sm} & \phi \\ I_m & 0 \end{bmatrix}}_{[\Phi]}\begin{bmatrix} q_m \\ \eta_{s_1} \end{bmatrix} \qquad [2.8]$$

where$\left\{\eta_{s_1}\right\}$ is the vector of generalized modal coordinates.

After substituting the transformation equation in [2.8] into equation [2.4], we obtain the system relatively to the considered substructure of dimension $m^{(k)} + s_1^{(k)}$:

$$\left[\tilde{K} - w^2\,\tilde{M}^{(k)}\right]\{\varsigma\}^{(k)} = \{\tilde{Q}\}^{(k)} \qquad [2.9]$$

where:

$$\left[\tilde{K}\right]^{(k)} = {}^t[\Phi][K]^{(k)}[\Phi] = \begin{bmatrix} \tilde{K}_{mm} & 0 \\ 0 & \Sigma \end{bmatrix}^{(k)} \qquad [2.10a]$$

$$\left[\tilde{K}_{mm}\right] = \left[K_{mm}\right] - {}^t\left[K_{sm}\right]\left[K_{ss}\right]^{-1}\left[K_{sm}\right] \qquad [2.10b]$$

$$[\Sigma] = {}^t[\phi][K_{ss}][\phi] = diag\left(\omega_1^2 \cdots \omega_p^2\right) \qquad [2.10c]$$

$$\left[\tilde{M}\right]^{(k)} = {}^{t}[\Phi][M]^{(k)}[\Phi] = \begin{bmatrix} \tilde{M}_{mm} & \tilde{M}_{mp} \\ \tilde{M}_{pm} & I_{p} \end{bmatrix}^{(k)} \qquad [2.10d]$$

$$\left[\tilde{M}_{mm}\right] = {}^{t}[\phi][M]^{(k)}[\phi] = [M_{mm}] - [M_{ms}][K_{ss}]^{-1}[K_{sm}] - [K_{ms}] \\ [K_{ss}]^{-1}[M_{sm}] + [M_{ms}][K_{ss}]^{-1}[M_{ss}][K_{ss}]^{-1}[K_{sm}] \qquad [2.10e]$$

$$\left[\tilde{M}_{pm}\right] = {}^{t}\left[\Phi_{s}\right][M]\left[\Phi_{s}\right] = [\phi]\left(\left[M_{sm}\right] - \left[M_{ss}\right]\left[K_{ss}\right]^{-1}\left[K_{sm}\right]\right) = {}^{t}\left[\tilde{M}_{mp}\right] \qquad [2.10f]$$

2.2.4. *Reduction of junction degrees of freedom*

One of the problems that face fixed-interface or free-interface multiplication methods is that all of the junction d.o.f. between the substructures are preserved. This significantly increases the size of the condensed model. Dynamic multiplication methods are sometimes not enough to reduce the order of the final condensed system, and the condensation of the junction d.o.f. becomes necessary [BRA 96; BO 92; TRA 01; SAR 04].

Reconsider the modal equation [2.4] of the motion of the complete system:

$$\left(\begin{bmatrix} K_{ss} & K_{sm} \\ K_{ms} & K_{mm} \end{bmatrix} - w^{2}\begin{bmatrix} M_{ss} & M_{sm} \\ M_{ms} & M_{mm} \end{bmatrix}\right)\begin{bmatrix} q_{s} \\ q_{m} \end{bmatrix} = \begin{bmatrix} 0 \\ Q_{m} \end{bmatrix}$$

Expanding the equations relating to the first group of rows yields a relation between $\{q_m\}$ and $\{q_s\}$, such as:

$$\{q_{s}\} = \left[t_{sm}(\lambda)\right]\{q_{m}\} \qquad [2.11]$$

with:

$$\left[t_{sm}\right] = -K_{sm}^{-1}K_{sm} \qquad [2.12a]$$

The method proposed by Guyan [GUY 65] is widely used in dynamic analysis of structures. It uses only the following static transformation $(\lambda = 0)$:

$$\left[t_{sm}\right] = -K_{sm}^{-1}K_{sm} \qquad [2.12b]$$

The transformation equation is expressed only in terms of junction d.o.f. (masters):

$$\{q\} = \begin{Bmatrix} q_s \\ \\ q_{\grave{u}} \end{Bmatrix} = \begin{bmatrix} t_{sm} \\ \\ I_m \end{bmatrix} \{q_m\} = \left[T_{sm}\right]\left[q_m\right] \qquad [2.13]$$

This transformation applied to the initial problem of the overall structure makes it possible to obtain a condensed problem as:

$$\left(\left[K_{cm}\right] - \lambda\left[M_{cm}\right]\right)\ \{q_m\} = 0 \qquad [2.14]$$

with

$$\left[K_{cm}\right] = {}^t\left[T_{sm}\right]\left[K\right]\left[T_{sm}\right]\left[M_{cm}\right] = \left[T_{sm}\right]\left[M\right]\left[T_{sm}\right] \qquad [2.15]$$

The displacement vector of the interface can be expressed by:

$$\left[q_m\right] = \left[\phi_m\right]\{\eta_m\} \qquad [2.16]$$

$\left[\phi_m\right]$ is the truncated matrix of the interface eigenmodes.

The displacement vector of the assembled structure becomes:

$$\{y_c\} = \begin{Bmatrix} \eta_p^{(1)} \\ \vdots \\ \eta_p^{(k)} \\ y_m \end{Bmatrix} = \underbrace{\begin{bmatrix} I^{(1)} & \cdots & 0 & 0 \\ \vdots & \ddots & & \\ 0 & \cdots & I^{(k)} & \\ 0 & \cdots & \cdots & \phi_m \end{bmatrix}}_{[T]} \begin{Bmatrix} \eta_p^{(1)} \\ \vdots \\ \eta_p^{(k)} \\ \eta_m \end{Bmatrix} \qquad [2.17]$$

In this case, the matrix for each substructure becomes:

$$[Z]^{(K)} = [Q]^{(K)}\,[\beta]^{(k)}\,[T] \qquad [2.18]$$

2.3. Coupling of modal synthesis and RBDO methods

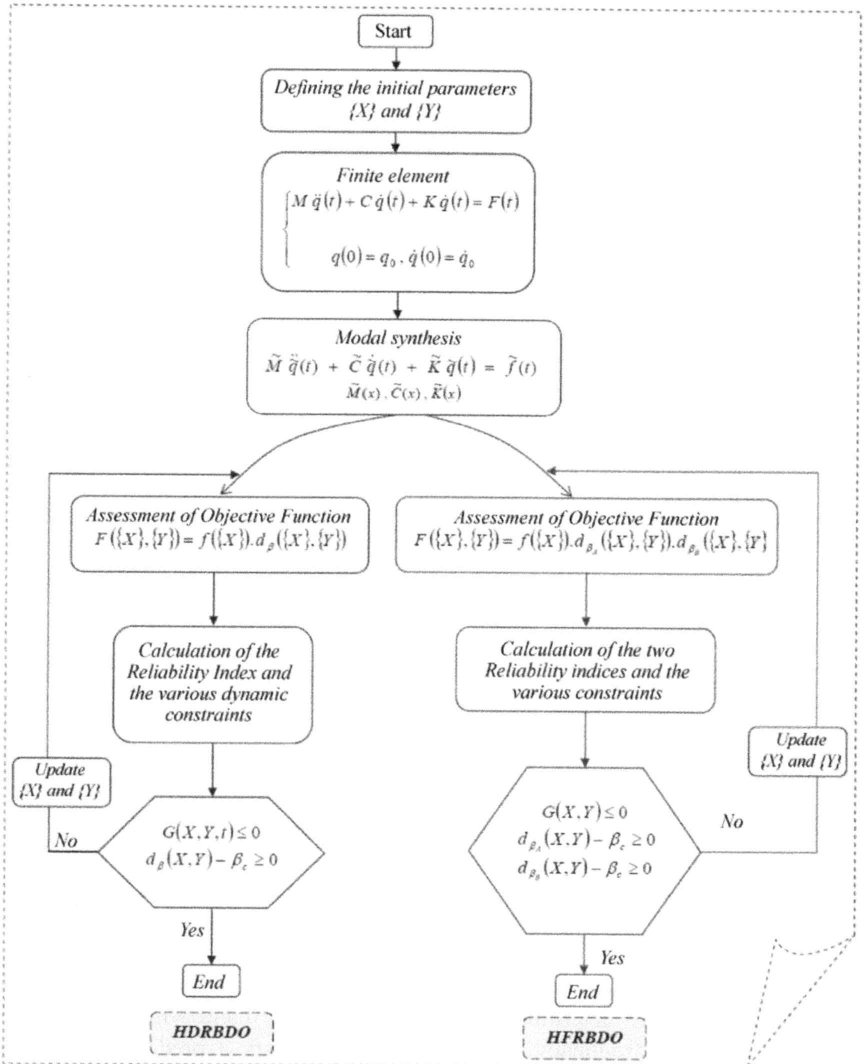

Figure 2.1. *RBDO coupling algorithm and model reduction. For a color version of this figure, see www.iste.co.uk/elhami/multi-physics.zip*

By applying RBDO procedures to real dynamic cases (for large-scale problems), the computation time becomes much higher compared to a small problem. The use of the sequential RBDO process can make the solution of real cases impossible because of its low stability and long computation time. However, the hybrid RBDO approach can effectively reduce computation time and allows coupling between different models. To increase its efficiency, we implement a coupling technique (see Figure 2.2) of the RBDO model and the reduction methods of the model [MOH 03]. The linkage process integrates several heterogeneous programs, and their communication is implemented through file transfers. First, a parametric finite model of an element is developed as a draft. The optimizer changes the design parameters to reach the optimal value of the objective function by satisfying all constraints. The last one is summarized into three steps.

Create a mesh of the model for given design variables and produce the matrices and boundary conditions on the displacements as well as the d.o.f.:

- using this information as input is formulating the scale model;
- the solution of the reduced system by the RBDO method;
- optimization stops or carries on according to convergence criteria.

2.4. Numerical application

2.4.1. *L-Plate embedded at the ends*

A plate ***L*** embedded at its two ends (see Figure 2.3) is considered for testing the effectiveness of the proposed approach. The geometric dimensions of the plate ***L*** are: the thicknesses ***R1*** and ***R2***, the length ***L1*** and the width ***L2*** (see Figure 2.2). The material used is steel with Young's modulus $E_0 = 2.110^{11}$ *N.m*, the density $\rho = 7800$ *kg/m*^{3} and Poisson's ratio $\nu = 0.3$.

The finite element model is generated by quadrilateral elements with six d.o.f. per node. It comprises 3822 d.o.f., of which 156 embedded d.o.f., that is, 36666 active ones.

The goal is to minimize the volume of the plate ***L*** that is subjected to the natural frequency stress and to determine the critical frequency band with respect to the targeted fundamental frequency. The system must satisfy a predetermined target reliability. We consider the target reliability index to be $\beta_c = 3.6$.

In this example, we compare the RBDO approach by direct computation (the full model) and with coupling of the different modal synthesis methods, the Craig Bampton (CB) method, the free-interface (IL) method, the Craig Bampton method with junction d.o.f. reduction (CBR) and the free-interface method with junction d.o.f. reduction (ILR) with respect to those of the full model.

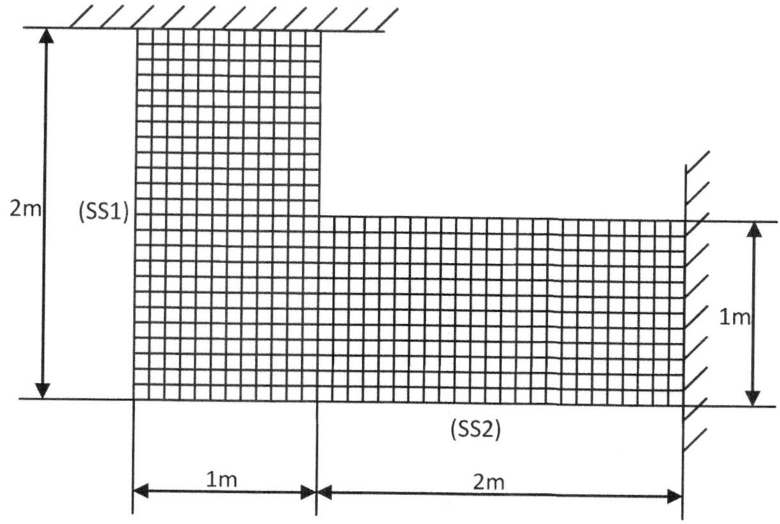

Figure 2.2. *Section of plate L. For a color version of this figure, see www.iste.co.uk/elhami/multi-physics.zip*

The structure is divided into two substructures, $SS^{(1)}$ and $SS^{(2)}$ (see Figure 2.3). Each substructure has 1950 d.o.f., of which junction d.o.f.

The strategy for choosing the number of modes (free or blocked) of each substructure involves taking into account all of the substructure modes contained in the multiple band of f_a that is twice the useful band $[0, 2f_u]$ [ELH 92].

The variables $R1$ and $R2$ are gathered into the random vector $\{Y\}$ having as averages the nominal values m_R and m_{R2}, grouped in the deterministic vector $\{x\}$, as standard deviations $0.1.m_x$.

We recall that the thickness of a free embedded plate according to Kirchhoff– Love's theory that the stiffness matrix depends on R^3 and the mass matrix depends on R.

The study will be carried out using four different methods in order to show the effectiveness and robustness of the proposed method. First, the computation is performed with a complete model of the structure. This computation will serve later as a reference calculation for calculations where the different methods of modal synthesis are used.

Parameter	*Design point (a)*	*Optimal solution*	*Design point (b)*
R1	*0.0095*	*0.0099*	*0.0103*
R2	*0.0081*	*0.0117*	*0.0152*
β	*3.6*	------	*3.6*
Frequency	***16.31***	***17.0000***	***17.69***
Volume	------	***0.0432***	-------

Table 2.1. *FRBDO results for the full model*

Table 2.1 presents the direct results of the hybrid frequency method (i.e. the full model). These results are found satisfying the various targeted constraints. There was a ***46.25%*** reduction in volume compared to the initial volume, and the critical frequency band is $[16.31, 17.69]$.

Parameter	*Design point (a)*	*Optimal solution*	*Design point (b)*
R1	*0.0121*	*0.0103*	*0.0082*
R2	*0.0063*	*0.0095*	*0.0124*
β	*3.6*	------	*3.6*
Frequency	***16.269***	***17.01***	***17.73***
Volume	------	***0.0388***	--------

Table 2.2. *RBDO results by integrating CB*

Parameter	*Design point (a)*	*Optimal solution*	*Design point (b)*
R1	*0.0121*	*0.0103*	*0.0082*
R2	*0.0063*	*0.0095*	*0.0124*
β	*3.6000*	-------	*3.6000*
Frequency	***16.269***	***17.0000***	***17.730***
Volume	-----	***0.0388***	-------

Table 2.3. *RBDO results using IL*

Tables 2.2 and 2.3 represent the computed results using modal synthesis methods with fixed and free interfaces without the reduction of junction d.o.f. The number of d.o.f. in the model is 96 d.o.f. and 104 d.o.f. respectively. The results for both methods satisfy and follow all of the required constraints. For both methods, we achieved the same ***51.5%*** reduction in volume compared to the initial volume.

Parameter	***Design point (a)***	***Optimal solution***	***Design point (b)***
R1	*0.0122*	*0.0103*	*0.0081*
R2	*0.0064*	*0.0094*	*0.0125*
β	*3.7*	-----	*3.6*
Frequency	***16.31***	***17.02***	***17.74***
Volume	--------	***0.0394***	-----

Table 2.4. *RBDO results using CBR*

Parameter	***Design point (a)***	***Optimal solution***	***Design point (b)***
R1	*0.0122*	*0.0103*	*0.0082*
R2	*0.0064*	*0.0094*	*0.0124*
β	*3.7*	-----	*3.6*
Frequency	***16.31***	***17.02***	***17.74***
Volume	--------	***0.0394***	-------

Table 2.5. *RBDO results using ILR*

Tables 2.4 and 2.5 represent the results computed by using the modal synthesis methods with fixed and free interface by taking into account the junction d.o.f. reduction. The number of d.o.f. in the reduced model is 37 d.o.f. and 49 d.o.f. respectively. As expected, we have a ***50.75%*** decrease in volume compared to the initial volume. The requested constraints were correctly satisfied.

2.4.2. *Analysis of results*

From the results obtained by the different methods used, we have a reduction of ***46.25%*** in volume compared to the initial volume for the direct method which was taken as a reference for the calculation. Computations by modal synthesis methods with fixed and free interfaces without the reduction of junction d.o.f. resulted in a ***10%*** decrease compared to the optimal volume found by the direct computation, and ***51.5%*** reduction in volume compared to the initial volume. However, the search for the critical frequency interval was provided with an error of ***0.24%*** compared to the frequency interval found by the direct method. The results found by the modal synthesis methods with fixed and free interfaces with a reduction in junction d.o.f. resulted in an ***8.8%*** decrease from the optimal volume found by the direct computation and a ***50.75%*** reduction in volume from the initial volume. However, the search for the critical frequency interval was provided with an error of ***0.14%*** compared to the frequency interval found by the direct method.

2.4.3. *A vibrating connecting rod*

The con rod connects the piston to the crankshaft. Together, they transform the linear motion (back and forth of the piston) into a rotary motion. The different geometric parameters and material properties are shown in Table 2.6.

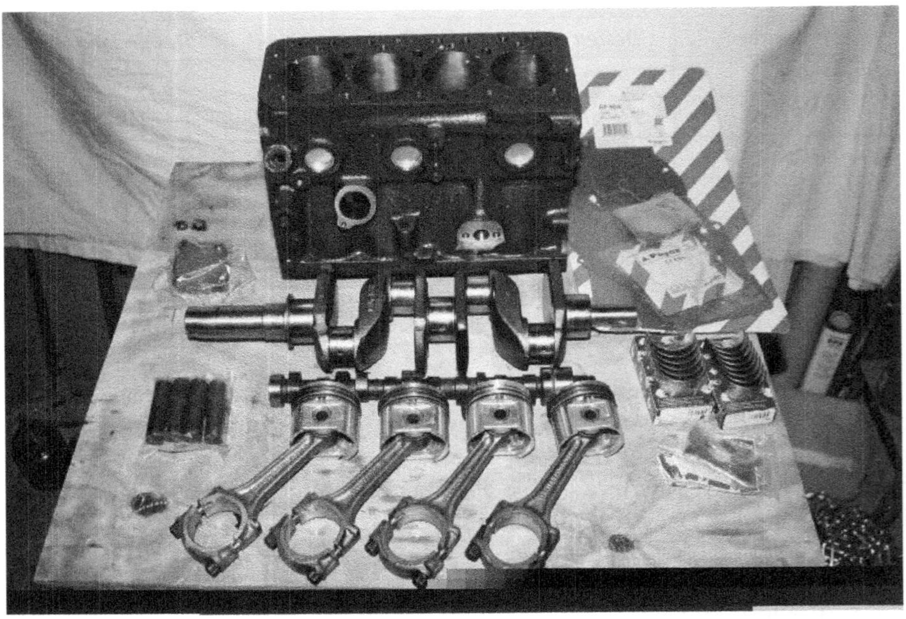

Figure 2.3. *Different connections for a connecting rod (inter). For a color version of this figure, see www.iste.co.uk/elhami/multi-physics.zip*

Parameter	*Young's modulus (N/mm^2)*	*Poisson's ratio*	*Material density (kg/mm^3)*
Numerical value	2.1×10^5	*0.3*	7800×10^{-9}

Table 2.6(a). *Material characteristics*

Parameter		*Numerical value*
L	*(mm)*	*150*
R1	*(mm)*	*28*
R2	*(mm)*	*12*
R3	*(mm)*	*6*
Rc1	*(mm)*	*5*
Rc2	*(mm)*	*5*
Rc3	*(mm)*	*4*
D3	*(mm)*	*10*
D5	*(mm)*	*6*
H1	*(mm)*	*15*
H2	*(mm)*	*10*
H3	*(mm)*	*5*

Table 2.6(b). *Geometric characteristics*

The objective is to determine by a probabilistic computation the critical region of the connecting rod at the level of the first resonance frequency. This study will be carried out by two methods: the first through a direct calculation of the RBDO hybrid frequency method and the second by inclusion of the modal synthesis method into the RBDO hybrid frequency approach. The predefined constraints are target reliability indices of $\beta_c = 3.8$ and a fundamental frequency of $f_c = 25Hz$.

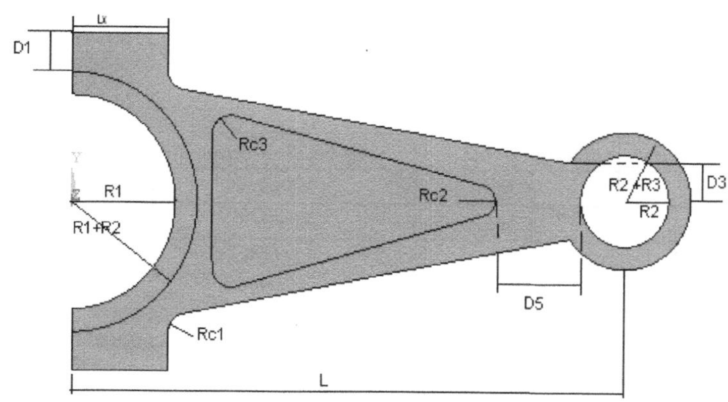

Figure 2.4(a). *Connecting rod dimensions in 2D. For a color version of this figure, see www.iste.co.uk/elhami/multi-physics.zip*

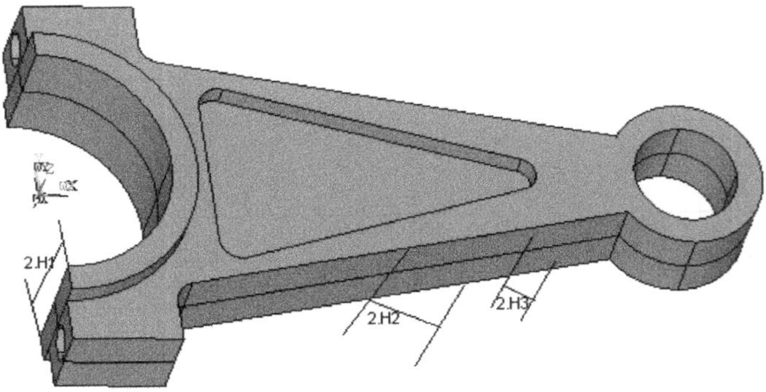

Figure 2.4(b). *Connecting rod dimensions in 3D. For a color version of this figure, see www.iste.co.uk/elhami/multi-physics.zip*

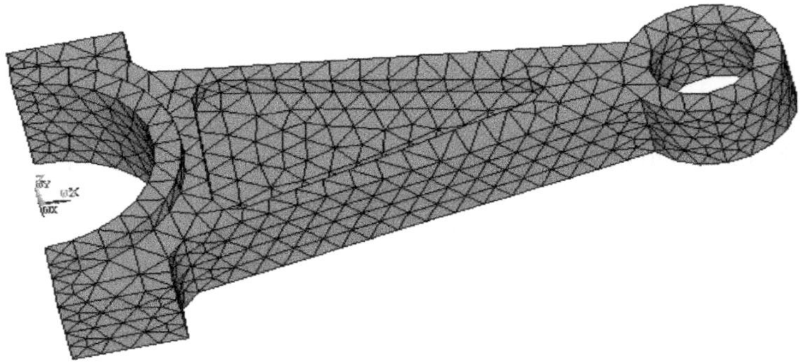

Figure 2.4(c). *The connecting rod mesh model. For a color version of this figure, see www.iste.co.uk/elhami/multi-physics.zip*

Variables	***Design point A***	***Optimal solution***	***Design point B***
R3	*5.2384*	*5.3638*	*5.8992*
D3	*8.2145*	*8.1267*	*8.7312*
D5	*4.6076*	*5.0318*	*4.1414*
H1	*13.288*	*15.975*	*14.737*
H2	*8.5483*	*8.0061*	*10.827*
H3	*3.4833*	*5.2535*	*5.9407*
Reliability index	*3.8*	--------	*3.8*
Frequency	*20.911*	*25.080*	*29.344*

Table 2.7. *Result of the full system RBDO hybrid frequency method*

Variables	*Design point A*	*Optimal solution*	*Design point B*
R3	*5.2846*	*5.3896*	*5.7938*
D3	*8.2220*	*8.3727*	*8.8042*
D5	*4.5676*	*5.0927*	*4.1224*
H1	*13.190*	*15.714*	*14.862*
H2	*8.5011*	*8.0071*	*10.838*
H3	*3.5124*	*5.2987*	*6.0829*
Reliability index	*3.8*	----	*3.82*
Frequency	*20.90*	*25.215*	*29.803*

Table 2.8. *Result of the RBDO method with the reduction of the degrees of freedom by the condensation method (1500 degrees of active freedom)*

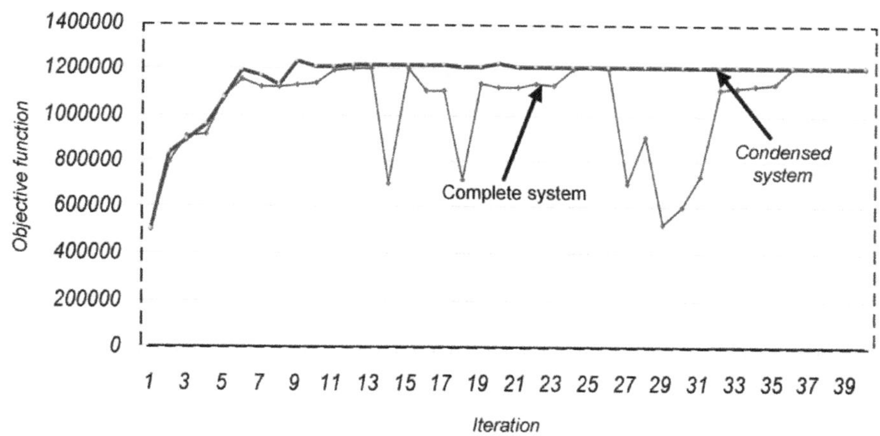

Figure 2.5. *Comparison of the convergence stability of the two objective functions. For a color version of this figure, see www.iste.co.uk/elhami/multi-physics.zip*

Table 2.7 presents the results of the RBDO hybrid frequency method for a full structure computation conserving all d.o.f. This result satisfied the various targeted constraints (we had a ***46.25%*** reduction in volume compared to the initial volume). Table 2.8 illustrates the computation made by the RBDO frequency hybrid approach with a condensation of the model including 1,500 degrees of active freedom. The result found satisfies the different targeted constraints (we had a ***0.5%*** reduction in volume compared to the optimal volume found before). The search for the critical frequency interval was provided with an error of ***0.24%*** compared to the frequency interval found by the direct method. However, including the modal synthesis methods has drastically reduced the number of d.o.f. and the computation time. Moreover, the convergence of the optimization problem is more stable compared to the direct computation (see Figure 2.5) and all of this confirms the quality and robustness of the proposed method.

2.4.4. *Application: transient model reduction*

This study will be carried out by two methods: a comparison of results found previously for a direct RBDO computation and the new results sought for by introducing the model into the dynamic hybrid approach.

The structure (see Figure 2.6) is modeled by eight-node hull elements. It comprises 14,000 degrees of active freedom. A condensation of the model is achieved in order to reduce the number of active d.o.f. to 1/3 of the initial d.o.f.

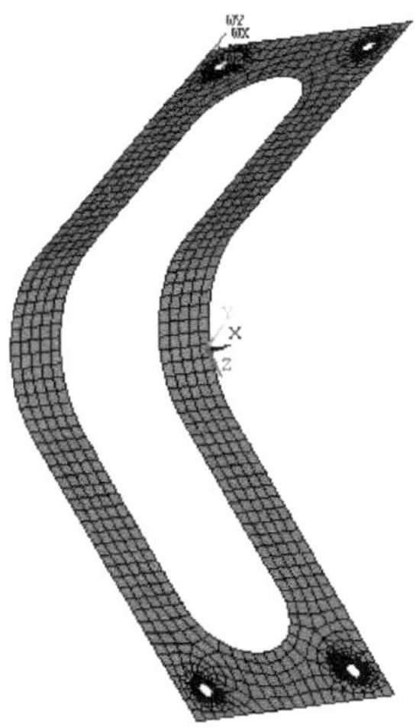

Figure 2.6. *The connecting rod mesh model. For a color version of this figure, see www.iste.co.uk/elhami/multi-physics.zip*

The different results are computed in Tables 2.8 and 2.9.

A comparison of the results calculated by the RBDO dynamic hybrid approach without model reduction and using the model reduction for the different probabilistic distributions: it was found that the different computations were satisfactory with respect to the different targeted constraints (Tables 2.8 and 2.9). With the use of the normal distribution, we had ***0.13%*** of the volume compared to the optimal volume found before and a total error with respect to the design parameters of ***2.27%***. However, the use of the log–normal distribution resulted in ***0.1%*** of the volume compared to the optimal volume found before and a total error with respect to the design parameters of ***0.01%***. Therefore, we had a ***7.6%*** reduction in volume for the most optimal volume with the log–normal distribution compared to the optimal volume with the normal distribution.

Parameter	***Full structure***		***Condensed structure***	
	Design point	*Optimal solution*	*Design point*	*Optimal solution*
L (in)	*1.8203*	*2.3425*	*1.8203*	*2.3425*
L2 (in)	*2.6456*	*2.9010*	*2.6462*	*2.9017*
R0 (in)	*0.14438*	*0.15876*	*0.14437*	*0.15888*
R1 (in)	*1.0137*	*1.0156*	*1.0137*	*1.0181*
σ_{max} *(psi)*	***36231***	***22843.15***	*36232.*	*22912.5452*
Reliability index	***3.8***	----	*3.81*	----
Volume (in^3)	-------	***4.23004718***	------	*4.22434903*

Table 2.9. *Results of the dynamic hybrid approach before and after reducing the model with random variables that follow a normal distribution*

Parameter	***Full structure***		***Condensed structure***	
	Design point	*Optimal solution*	*Design point*	*Optimal solution*
L (in)	*1.8201*	*2.3447*	*1.820*	*2.344*
L2 (in)	*2.6292*	*2.9343*	*2.629*	*2.934*
R0 (in)	*0.14460*	*0.14726*	*0.14460*	*0.14726*
R1 (in)	*1.0150*	*1.1573*	*1.0150*	*1.1573*
σ_{max} *(psi)*	***36231***	***27131.1439***	*36231.*	*27134.1405*
Reliability index	***3.8***	--------	*3.8*	------------
Volume (in^3)	-------	***3.9037641***	-----------	*3.90*

Table 2.10. *Results of the RBDO dynamic hybrid approach before and after reducing the model with random variables that follow a log–normal distribution*

2.5. Conclusion

In this chapter, we have presented the different techniques of modal synthesis methods. It should be mentioned that the conventional substructuring methods have a major disadvantage since the junction d.o.f. remain in the state vector of the final condensed system. To overcome this drawback, a method for reducing these junction d.o.f. based on the junction modes has been used. These modes are obtained from the condensation at the Guyan interfaces of the full structure. Then, modal synthesis techniques were integrated into the process of reliable design optimization. Through the various applications studied, we have shown the efficiency and robustness of the model.

3

Modal Synthesis Methods and Stochastic Finite Element Methods

3.1. Introduction

This chapter is a presentation of the general equations of structural dynamics with applications to stochastic modal synthesis methods. The first section is a reminder of the equations of motion for a structure. It sets the notations for the calculation of dynamic responses, the calculation of eigenmodes and frequency response functions. These problems will be addressed later in the text. The solution of the dynamic equilibrium equation system for a structure can be achieved using one of two conventional strategies [IMB 95]. The most common solution strategy in dynamics is modal superposition, which is suitable for linear structures where only the first eigenmodes are excited. On the other hand, the direct solution methods involve integrating the equations of motion, in order to address nonlinear structures. The reader can also refer to [MAN 13; ABI 19].

In the second part, the strategy of dynamic substructuring is recalled, initially formulated for static problems [PRZ 63]. It involves treating a structure as an assembly of substructures interconnected with one another. Modal synthesis methods differ in the choice of modes to represent the dynamics of each substructure and in the assembly procedures. The most well-known methods are the methods using fixed-interface substructural modes [CRA 68; HUR 65; HUR 71] and free-interface substructural modes [GOL 69; TOU 01]. Another alternative consists of using branches modes. These modes are obtained by imposing additional inertial or elastic loads at the boundaries [GLA 64; HRU 71; JEZ 85].

Moreover, when considering structures with complex geometries, the size of the final condensed system is often significant because of the high number of junction degrees of freedom (d.o.f.) kept for assembly. In the literature, methods are proposed to reduce the number of junction d.o.f. [BRA 96; BOU 92; TRA 01]. In this part, we propose a strategy for the reduction of the junction d.o.f., after assembly. This strategy is based on the use of interface modes. These modes are obtained from the Guyan condensation at the interfaces of the full structure.

The third section is dedicated to the description of the different tools for the calculation of the characteristics of structural responses with uncertain parameters. The use of these tools is conditioned by the modeling of the different sources of uncertainty that occur during a process. Monte Carlo simulations [SHI 72] are widely used, despite their high cost in terms of computational time, and serve as a reference for approximate calculations. The perturbation method [KLE 92], which is based on a low-order (usually one or two) Taylor series expansion, is the most widely used method, despite convergence problems. Neumann's series expansion is also used [YAM 88]. Another method, based on projections onto a basis of orthogonal polynomials with Gaussian variables [GHA 90] can be used.

3.2. Linear dynamical problems

3.2.1. *Equations of motion*

The finite element method applied in dynamics of a mechanical structure with N d.o.f. leads to the following system of equations:

$$[M]\{\ddot{y}(t)\}+[C]\{\dot{y}(t)\}+[K]\{y(t)\}=\{F(t)\} \qquad [3.1a]$$

where [M], [C] and [K] are the mass, damping and stiffness matrices, respectively. They are Nth-order squares. The matrix [M] is symmetric definite positive; the matrix [K] is symmetric definite nonnegative.

The damping matrix is not computable in practice. Typically, damping is introduced into models in the form of modal damping. In this case, the system of equations of motion in equation [3.1a] is reduced to decoupled second-order differential equations:

$$\ddot{\eta}_i(t)+2\xi_i\omega_i\dot{\eta}_i(t)+\omega_i^2\,\eta_i(t)=f_i(t) \qquad [3.1b]$$

where ω_i is the i-th natural angular frequency of the structure, ξ_i is the damping factor of the i-th mode and f_i is the generalized force of the i-th mode.

3.2.2. *Solution in transient state*

The solution $\eta_i(t)$ to the previous classical differential equations can be written as convolution products of the excitation function and the impulse response $h_i(t)$ of mode i [ELH 92]:

$$\eta_i(t) = \int_0^t h_i(t-\tau) f_i(\tau) d\tau + \eta_i(0) g_i(t) + \dot{\eta}_i(0) h_i(t) \quad [3.2]$$

If mode i has a subcritical damping $(\xi_i < 1)$, we have:

$$h_i(t) = \frac{e^{-\xi_i \omega_i t} \sin\sqrt{1-\xi_i^2}\,\omega_i t}{\sqrt{1-\xi_i^2}\,\omega_i} \quad [3.3a]$$

$$g_i(t) = e^{-\xi_i \omega_i t} \left[\cos\sqrt{1-\xi_i^2}\,\omega_i t + \frac{\xi_i \sin\sqrt{1-\xi_i^2}\,\omega_i t}{\sqrt{1-\xi_i^2}} \right] \quad [3.3b]$$

For critical damping, $(\xi_i = 1)$, we have:

$$h_i(t) = t\, e^{-\omega_i t} \quad [3.4a]$$

$$g_i(t) = (1-\omega_i t) e^{-\omega_i t} \quad [3.4b]$$

For supercritical damping $(\xi_i > 1)$, we have:

$$h_i(t) = \frac{e^{-\xi_i \omega_i t}\, sh\sqrt{\xi_i^2 - 1}\,\omega_i t}{\sqrt{\xi_i^2 - 1}\,\omega_i} \quad [3.5a]$$

$$g_i(t) = e^{-\xi_i \omega_i t} \left[ch\sqrt{\xi_i^2 - 1}\,\omega_i t + \frac{\xi_i\, sh\sqrt{\xi_i^2 - 1}\,\omega_i t}{\sqrt{\xi_i^2 - 1}} \right] \quad [3.5b]$$

The Duhamel integral is then numerically evaluated.

3.2.3. *Solution in the harmonic domain*

Consider the case of a sinusoidal excitation with angular frequency ω, we have:

$$\{F(t)\} = \{F\}e^{i\omega t} \qquad [3.6]$$

The harmonic solution will therefore be:

$$\{y(t)\} = \{y\}e^{i\omega t} \qquad [3.7]$$

The equation of motion of equation [3.13] becomes:

$$(-\omega^2[M] + i\omega[C] + [K])\{y\} = \{F\} \qquad [3.8]$$

The transfer function $H_{ij}(\omega)$ is defined as:

$$y_i = H_{ij}\, F_j \qquad [3.9]$$

The frequency response function $H_{ij}(\omega)$ may not exist for some values of ω (in the presence of instability, undamped resonance, etc.).

3.2.3.1. *Direct computation*

The direct computation consists of inverting, for each angular frequency value sought for, the matrix system:

$$[H] = (-\omega^2[M] + i\omega[C] + [K])^{-1} \qquad [3.10]$$

When the size of the matrices is significant, this type of computation can become very expensive. Another method involves using the eigenmodes of the structure.

3.2.3.2. *Calculation by modal superposition*

Using the truncated modal basis (m < n), the following fundamental expression is obtained:

$$\{y(t)\} = \sum_{i=1}^{m} \{\phi_i\}\eta_i(t) = [\varphi]\{\eta\} \qquad [3.11]$$

By pre-multiplying by ${}^{t}\{\phi_j\}$, using the orthogonality properties of the eigenmodes, we get the modal contributions:

$$y_j = \frac{{}^{t}\{\phi_j\}\{F\}}{\omega_j^2 - \omega^2 + 2i\xi_j \omega \omega_j} \tag{3.12}$$

The transfer function is then given by:

$$H_{ij}(\omega) = \sum_{m=1}^{M} \frac{\{\phi_m\}_i \{\phi_m\}_j}{\omega_m^2 - \omega^2 + 2i\xi_m \omega \omega_m} \tag{3.13}$$

3.2.4. *Direct integration*

There are many methods for integrating the equations of motion. The general procedure is to discretize time and to formulate what is happening at the time $t + \Delta t$ as a function of what is happening at time t based on limited expansions of displacements, velocities and/or acceleration, more or less similar to Taylor's expansions. The main question about these methods is their stability with respect to their accuracy. Let us consider that we are looking for the solution to the equation between time t = 0 and time T, the time interval will be divided into n intervals of the same length $\Delta t = \frac{T}{n}$. We will seek for the vectors y(t), $y(t)$, $\ddot{y}(t)$ at instants Δt,2 Δt,....,t,t + Δt,...., T, knowing that the initial conditions are given.

3.2.4.1. *Newmark method*

The expansions of the displacements and velocities are given by [ELH 92]:

$$\{\dot{y}(t+\Delta t)\} = \{\dot{y}(t)\} + [(1-\delta)\{\ddot{y}(t)\} + \delta\{\ddot{y}(t+\Delta t)\}]\Delta t \tag{3.14a}$$

$$\{y(t+\Delta t)\} = \{y(t)\} + \{\dot{y}(t)\}\Delta t + [(0.5-\alpha)\{\ddot{y}(t)\} + \alpha\{\ddot{y}(t+\Delta t)\}](\Delta t)^2 \tag{3.14b}$$

in which α and δ are the two parameters that ensure the accuracy and stability of the solution. When $\delta \geq 0.5$ and $\alpha \geq (\delta + 0.5)/4$, the Newmark method is unconditionally stable.

From equations [3.14a] and [3.14b], $\{\dot{y}(t+\Delta t)\}$ and $\{\ddot{y}(t+\Delta t)\}$ can be found as a function only of the unknown $\{y(t+\Delta t)\}$ and the solution at time t:

$$\{\ddot{y}(t+\Delta t)\} = a_0[\{y(t+\Delta t)\} - \{y(t)\}] - a_2\{\dot{y}(t)\} - a_3\{\ddot{y}(t)\} \qquad [3.15a]$$

$$\{\dot{y}(t+\Delta t)\} = \{\dot{y}(t)\} + a_6\{\ddot{y}(t)\} + a_7\{\ddot{y}(t+\Delta t)\}$$

in which:

$$a_0 = \frac{1}{\alpha\Delta t^2} \quad a_1 = \frac{\delta}{\alpha\Delta t} \quad a_2 = \frac{1}{\alpha\Delta t} \quad a_3 = \frac{1}{2\alpha} - 1 \quad a_4 = \frac{\delta}{\alpha} - 1 \quad a_5 = \frac{\Delta t}{2}(\frac{\delta}{\alpha} - 2)$$

$$a_6 = \Delta t(1-\delta) \quad a_7 = \delta\Delta t \qquad [3.15b]$$

The equation of motion [3.1] is given at time $t + \Delta t$ by:

$$[M]\{\ddot{y}(t+\Delta t)\} + [C]\{\dot{y}(t+\Delta t)\} + [K]\{y(t+\Delta t)\} = \{F(t+\Delta t)\} \qquad [3.16]$$

Replacing equations [3.15a] and [3.15b] in equation [3.16], the following equations are obtained:

$$[K_{eq}]\{y(t+\Delta t)\} = \{F_{eq}\} \qquad [3.17a]$$

with

$$[K_{eq}] = [K] + a_0[M] + a_1[C] \qquad [3.17b]$$

$$\begin{aligned}\{F_{eq}\} = \{F(t+\Delta t)\} &+ [M](a_0\{y(t)\} + a_2\{\dot{y}(t)\} + a_3\{\ddot{y}(t)\}) \\ &+ [C](a_1\{y(t)\} + a_4\{\dot{y}(t)\} + a_5\{\ddot{y}(t)\})\end{aligned} \qquad [3.17c]$$

Once the displacement $\{y(t+\Delta t)\}$ has been determined from equation [3.17a], the velocities $\{\dot{y}(t+\Delta t)\}$ and acceleration $\{\ddot{y}(t+\Delta t)\}$ will be obtained from equations [3.15a] and [3.15b].

3.2.4.2. *θ-Wilson method*

This is an extension of a method in which the acceleration is assumed to vary linearly over the interval $[n\Delta t,(n+1)\Delta t]$; Wilson [KLE 92] assumes that this linear variation occurs over the interval $[n\Delta t,(n+\theta)\Delta t]$. θ is a parameter (the value 3.4 is recommended by Wilson). If τ denotes the time in the interval $[0, \theta\Delta t]$, then the acceleration in the interval $[t, t+\theta\Delta t]$ is written as:

$$\{\ddot{y}(t+\tau)\} = \{\ddot{y}(t)\} + \frac{\tau}{\theta\Delta t}\left(\{\ddot{y}(t+\theta\Delta t)\} - \{\ddot{y}(t)\}\right) \qquad [3.18]$$

Speed and displacement are obtained by successive integrations:

$$\{\dot{y}(t+\tau)\} = \{\dot{y}(t)\} + \tau\{\ddot{y}(t)\} + \frac{t^2}{2\theta\Delta t}\left(\ddot{y}(t+\theta\Delta t) - \{\ddot{y}(t)\}\right) \qquad [3.19a]$$

$$\{y(t+\tau)\} = \{y(t)\} + \tau\{\dot{y}(t)\} + \frac{1}{2}\tau^2\{\ddot{y}(t)\} + \frac{t^3}{6\theta\Delta t}\left(\{\ddot{y}(t+\theta\Delta t)\} - \{\ddot{y}(t)\}\right) \qquad [3.19b]$$

Let $t = \theta\Delta t$, the equations enable us to determine $\{\ddot{y}(t+\theta\Delta t)\}$ and $\{\dot{y}(t+\theta\Delta t)\}$ are:

$$\{\ddot{y}(t+\theta\Delta t)\} = \frac{6}{(\theta\Delta t)^2}\left(\{y(t+\theta\Delta t)\} - \{y(t)\}\right) - \frac{6}{\theta\Delta t}\{\dot{y}(t)\} - 2\{\ddot{y}(t)\} \qquad [3.20a]$$

$$\{\dot{y}(t+\theta\Delta t)\} = \frac{3}{\theta\Delta t}\left(\{y(t+\theta\Delta t)\} - \{y(t)\}\right) - 2\{\dot{y}(t)\} - \frac{\theta\Delta t}{2}\{\ddot{y}(t)\} \qquad [3.20b]$$

By replacing equations [3.19a] and [3.19b] in the equation of motion of [3.1] at time $t+\theta\Delta t$, the following equations are obtained to determine $\{y(t+\theta\Delta t)\}$:

$$\left[K_{eq}\right]\{y(t+\theta\Delta t)\} = \{F_{eq}\} \qquad [3.21a]$$

with

$$\left[K_{eq}\right] = [K] + \frac{6}{(\theta\Delta t)^2}[M] + \frac{3}{\theta\Delta t}[C] \qquad [3.21b]$$

$$\{F_{eq}\} = \{F(t)\} + \theta\left[\{F(t+\Delta t)\} - \{F(t)\}\right] + [M]$$
$$\left(\frac{6}{(\theta\Delta t)^2}\{y(t)\} + \frac{6}{\theta\Delta t}\{y(t)\} + 2\{y(t)\}\right) \quad [3.21c]$$
$$+[C]\left(\frac{3}{\theta\Delta t}\{y(t)\} + 2\{\dot{y}(t)\} + \frac{\theta\Delta t}{2}\{\ddot{y}(t)\}\right)$$

Once the displacement $\{y(t+\theta\Delta t)\}$ is determined from equation [3.21a], the velocities $\{\dot{y}(t+\theta\Delta t)\}$ and acceleration $\{\ddot{y}(t+\theta\Delta t)\}$ vectors will be obtained from equations [3.20a] and [3.20b].

3.3. Modal synthesis methods

3.3.1. *Introduction*

One of the challenges existing in structural dynamics is to determine a sufficient number of frequencies and eigenmodes, as they are the basis for predicting vibrational behavior. To achieve this objective, methods based on a discretization of the structure under consideration have been developed. However, the accuracy necessary to describe an industrial structure generally requires a model comprising a large number of elements. Consequently, for large structures, the result is a significant model size. To overcome this kind of problem, dynamic substructuring methods [HUR 65; CRA 68; MAC 71; RUB 75] were developed, considered to be the best strategies for the dynamic analysis of complex structures. The overall structure is divided into several substructures $SS^{(s)}$ (s=1,..., S) linked by their interfaces. S is the number of substructures. Each of the substructures is then separately condensed using a reduced number of parameters. From these characteristics, a system of a much smaller size than that of the original system corresponding to the full structure is obtained. All matrices are assigned the upper subscript (s) relative to the substructure $SS^{(s)}$. The displacement vector for each substructure is as follows:

$$\{y\}^{(s)} = \begin{Bmatrix} y_i \\ y_j \end{Bmatrix}^{(s)} \quad [3.22a]$$

$\{y_j\}^{(s)}$ (jx1) is the subvector of the junction d.o.f. and $\{y_i\}^{(s)}$ (ix1) is the subvector of the internal d.o.f.

Similarly, for the vector of applied forces, we get:

$$\{f\}^{(s)} = \{f_j\}^{(s)} + \{f_e\}^{(s)} \qquad [3.22b]$$

$\{f_j\}^{(s)}$ is the force vector at the interface and $\{f_e\}^{(s)}$ is the external force vector.

We have the corresponding partition of the stiffness and mass matrices:

$$[K] = \begin{bmatrix} K_{ii} & K_{ij} \\ K_{ji} & K_{jj} \end{bmatrix} \quad [M] = \begin{bmatrix} M_{ii} & M_{ij} \\ M_{ji} & M_{jj} \end{bmatrix} \qquad [3.23]$$

The principle of substructuring is based on the choice of representation modes of the displacements of each substructure.

The displacement vector $\{y\}$ of each substructure can be expressed as:

$$\{y\}^{(s)} = [Q]^{(s)} \begin{Bmatrix} y_j^{(s)} \\ \eta_p^{(s)} \end{Bmatrix} = [Q]^{(s)} \{y_c\}^{(s)} \qquad [3.24]$$

$\{\eta_p\}^{(s)}$ is the vector of generalized coordinates.

The matrix [Q] will be defined according to the substructuring method used.

NOTE.– This formulation cannot be applied to the Mac Neal method presented below. Using equation [3.24], the kinetic energy of each substructure $SS^{(s)}$ becomes:

$$T^{(s)} = \frac{1}{2}\, {}^t\{\dot{y}_c\}^{(s)} \left[M_c\right]^{(s)} \{\dot{y}_c\}^{(s)} \qquad [3.25a]$$

with

$$[M_c]^{(s)} = {}^t[Q]^{(s)} [M]^{(s)} [Q]^{(s)} \qquad [3.25b]$$

Similarly, for the potential energy of each substructure $SS^{(s)}$ is:

$$U^{(s)} = \frac{1}{2}\, {}^t\{y_c\}^{(s)} \left[K_c\right]^{(s)} \{y_c\}^{(s)} \qquad [3.26a]$$

with

$$\left[K_c\right]^{(s)} = {}^t\left[\mathrm{Q}\right]^{(s)}\left[\mathrm{K}\right]^{(s)}\left[\mathrm{Q}\right]^{(s)} \qquad [3.26b]$$

3.3.2. *Substructure assembly technique*

The equation of motion of the assembled structure is obtained by expressing the compatibility conditions at the interface:

– Continuity of movement is written as:

$$\left\{\mathrm{y_j}\right\}^{(1)} = \left\{\mathrm{y_j}\right\}^{(2)} = \cdots = \left\{\mathrm{y_j}\right\}^{(S)} = \left\{\mathrm{y_j}\right\} \qquad [3.27a]$$

– The equilibrium of the interface:

$$\sum_{s=1}^{S} \left\{\mathrm{f_j}\right\}^{(s)} = 0 \qquad [3.27b]$$

Two approaches can be used:

1) Assembly by the method of Lagrange equations with constraints [ELH 96]:

The equation of motion of the assembled structure is obtained from the Lagrange equations with multipliers introducing the bonding relations.

2) Superelement assembly method [ELH 92]:

The conservation of junction d.o.f. in dynamic substructuring methods makes it possible to assemble these matrices as ordinary finite elements.

Let $\{y_c\}$ be the vector of displacements independent of the assembled structure:

$$\left\{\mathrm{y}_c\right\} = \begin{Bmatrix} \eta_p^{(1)} \\ \vdots \\ \eta_p^{(S)} \\ \mathrm{y}_j \end{Bmatrix} \qquad [3.28]$$

The compatibility of the junction displacements of the assembled structure is obtained by writing for each substructure $SS^{(s)}$ the following matrix relation:

$$\{y_c\}^{(s)} = [\beta]^{(s)} \{y_c\} \qquad [3.29]$$

$[\beta]^{(s)}$ is the geometry localization or connectivity matrix of the substructure $SS^{(s)}$. It serves as the basis for localizing or identifying the d.o.f. of each substructure $SS^{(s)}$ in all of the d.o.f. of the assembled structure. $[\beta]^{(s)}$ are the Boolean matrices whose elements are 0s or 3s.

A transformation matrix is then defined for each substructure:

$$[Z]^{(s)} = [Q]^{(s)} [\beta]^{(s)} \qquad [3.30]$$

The condensed mass and stiffness matrices are then given by:

$$[M_c] = \sum_{s=1}^{S} {}^t[Z]^{(s)} [M]^{(s)} [Z]^{(s)} \qquad [3.31]$$

$$[K_c] = \sum_{s=1}^{S} {}^t[Z]^{(s)} [K]^{(s)} [Z]^{(s)} \qquad [3.32]$$

3.3.3. *Fixed-interface method*

The motion of each substructure can be characterized by its fixed-interface modes as well as the static modes of bonding [CRA 68]. The index (s) has been omitted in equations [3.33]–[3.39]:

$$\{y\} = [Y]\{\eta\} + [\psi_c]\{y_j\} \qquad [3.33]$$

In this case, matrix [Q] is then given by

$$[Q] = [\psi_c \quad Y] \qquad [3.34]$$

[Y] is the truncated modal basis of the fixed-interface substructure, that is, $\{y_j\} = \{0\}$ it can be written as:

$$[Y] = \begin{bmatrix} \varphi_{ip} \\ 0 \end{bmatrix} \tag{3.35a}$$

with

$$\left[\varphi_{ip}\right] = [\{x_i\}^{(1)} \ldots\ldots\ldots \{x_i\}^{(p)} \ldots\ldots\ldots] \tag{3.35b}$$

such that $x_i^{(p)}$ as the eigen solution of the homogeneous system:

$$[K_{ii}]\{x_i\} = \lambda[M_{ii}]\{x_i\} \tag{3.35c}$$

$[\psi_c]$ is the matrix of static modes of connection obtained by calculating the displacements corresponding to a unit displacement in one junction d.o.f., assuming that the other junction d.o.f. are fixed, obtained by:

$$\begin{bmatrix} \mathrm{K}_{ii} & \mathrm{K}_{ij} \\ \mathrm{K}_{ji} & \mathrm{K}_{jj} \end{bmatrix} \begin{bmatrix} \phi_{ij} \\ I_j \end{bmatrix} = \begin{bmatrix} 0 \\ F_j \end{bmatrix} \tag{3.36}$$

Expanding the equations for the first group of rows leads to:

$$\phi_{ij} = - K_{ii}^{-1} K_{ij} \tag{3.37a}$$

The matrix $[\psi_c]$ then becomes:

$$[\psi_c] = \begin{bmatrix} - K_{ii}^{-1} K_{ij} \\ I_j \end{bmatrix} \tag{3.37b}$$

The condensed mass matrix for each substructure defined in [3.26b]:

$$[M_c] = \begin{bmatrix} I_p & \bar{M}_{pj} \\ \bar{M}_{jp} & \bar{M}_{jj} \end{bmatrix} \tag{3.38a}$$

An expression in which the following submatrices are found:

– Condensed mass matrix at interfaces:

$$\bar{M}_{jj} = {}^{t}[\psi_c][M][\psi_c] \qquad [3.38b]$$

– Mass coupling matrix between static interface modes and fixed-boundary eigenmodes

$$\bar{M}_{jp} = {}^{t}[\psi_c][M][Y] \qquad [3.38c]$$

Similarly for the condensed stiffness matrix for each substructure defined in [3.27b]:

$$[K_c] = \begin{bmatrix} \Omega_p & 0 \\ 0 & \bar{K}_{jj} \end{bmatrix} \qquad [3.39a]$$

An expression in which the following submatrices are found:

– Rigidity matrix statically condensed to interface d.o.f.

$$\bar{K}_{jj} = {}^{t}[\psi_c][K][\psi_c] = K_{jj} - {}^{t}K_{ij}\, K_{ii}^{-1}\, K_{ij} \qquad [3.39b]$$

– Diagonal matrix of generalized rigidities of embedded eigenmodes

$$[\Omega_p] = {}^{t}\varphi_{ip}\, K_{ii}\, \varphi_{ip} \qquad [3.39c]$$

To achieve the assembly, we define for each substructure(s) a matrix of the location of the junction d.o.f. such that:

$$\{y_j\}^{(s)} = [\beta_j]^{(s)}\{y_j\} \qquad [3.40]$$

The matrices $[M_c]$ and $[K_c]$ defined in equations [3.33a] and [3.33b] are given in this case by:

$$\left[M_c\right]=\begin{bmatrix} I_p^{(1)} & & & \bar{M}_{pj}^{(1)} \\ & \ddots & & \\ & & I_p^{(S)} & \bar{M}_{pj}^{(S)} \\ & & & \bar{M}_{jj} \end{bmatrix} \quad \left[K_c\right]=\begin{bmatrix} \Omega_p^{(1)} & & & \\ & \ddots & & \\ & & \Omega_p^{(S)} & \\ & & & \bar{K}_{jj} \end{bmatrix} \qquad [3.41a]$$

with:

$$\bar{M}_{jj}=\sum_{s=1}^{S} {}^t\left[\beta_j\right]^{(s)} \bar{M}_{jj}^{(s)} \left[\beta_j\right]^{(s)} \qquad [3.41b]$$

$$\bar{M}_{jp}^{(s)}={}^t\left[\beta_j\right]^{(s)} \bar{M}_{jp}^{(s)} \qquad [3.41c]$$

$$\bar{K}_{jj}=\sum_{s=1}^{S} {}^t\left[\beta_j\right]^{(s)} \bar{K}_{jj}^{(s)} \left[\beta_j\right]^{(s)} \qquad [3.41d]$$

Let $\left[Y\right]$ and $\{\eta\}$ be the modal matrix and the vector of modal coordinates corresponding to the eigenvalue problem of the full system, that is,

$$\begin{bmatrix} \eta_p^{(1)} \\ \vdots \\ \eta_p^{(S)} \\ y_j \end{bmatrix}=\begin{bmatrix} Y_p^{(1)} \\ \vdots \\ Y_p^{(S)} \\ Y_j \end{bmatrix}\{\eta\} \qquad [3.42]$$

Hence, for each substructure, using relation [3.34], the following substitution relations are obtained:

$$\left\{y_j\right\}^{(s)}=\left[\beta_j\right]^{(s)}\left[Y_j\right]\{\eta\} \qquad [3.43a]$$

$$\left\{\eta_p\right\}^{(s)}=\left[Y_p\right]^{(s)}\{\eta\} \qquad [3.43b]$$

$$\left\{y_i\right\}^{(s)}=\left[\phi_{ij}^{(s)}\,\beta_j^{(s)} Y_j+\varphi_{ip}^{(s)} Y_p^{(s)}\right]\{\eta\} \qquad [3.43c]$$

The main advantage of this method resides in its accuracy. Its formulation is fully compatible with the organization of structural analysis software. The disadvantages are as follows:

– In the system state vector, interface d.o.f. can be elevated in the case of complex structures.

– This method is virtually incompatible with the use of substructure data of experimental origin [IMB 95].

3.3.4. *MacNeal's open-interface method*

The most common free-interface modal synthesis method is that by Mac Neal [MAC 71].

Let the modal equation of the motion of each substructure(s) taken separately without damping:

$$([K]-\lambda[M])\{y\}=\{f\} \quad [3.44]$$

The movement of each substructure can be characterized by its free-interface modes, that is, for interface d.o.f.:

$$\{y\} = [Y]\{\eta\} \quad [3.45a]$$

where [Y] is the truncated modal basis of the free-interface substructure.

The use of a truncated modal matrix can induce unacceptable errors on the dynamic substructural behavior. This drives forward the need to take into account the static effect of upper modes truncated by a residual flexibility correction. Hence, the basic relation of each substructure:

$$\{y\} = [Y]\{\eta\} + [R]\{f\} \quad [3.45b]$$

where $[R]$ is the residual flexibility matrix. It represents the contribution of the methods that were not retained.

Given the orthogonality of matrices $[Y]$ and $[R]$ with respect to the matrices of mass [M] and stiffness [K]:

$$^t[Y][M][R]=[0] \quad ^t[Y][K][R]=[0] \quad [3.46]$$

The pre-multiplication by ${}^t[Y]$ and relation [3.46] make it possible to rewrite equation [3.44] as:

$$([\Omega]^{(s)} - \lambda[I]^{(s)})\,\{\eta\}^{(s)} = {}^t[Y]^{(s)}\{f\}^{(s)} \qquad [3.47]$$

The distribution of the vector $\{y\}$ in $\{y_i\}$ and $\{y_j\}$ previously described enables us to have the matrices $[\mathrm{Y}]$ and $[\mathrm{R}]$ in the following way:

$$[\mathrm{Y}]^{(s)} = \begin{bmatrix} \mathrm{Y}_{\mathrm{i}}^{(s)} \\ \mathrm{Y}_{\mathrm{j}}^{(s)} \end{bmatrix} \quad [\mathrm{R}]^{(s)} = \begin{bmatrix} R_{ii}^{(s)} & R_{ij}^{(s)} \\ {}^t R_{ij}^{(s)} & R_{jj}^{(s)} \end{bmatrix} \qquad [3.48]$$

Moreover, the modal equations of each substructural motion taken in isolation without damping and without external force $\left(\{f_e\}^{(s)} = 0\right)$ are written as:

$$([\Omega]^{(s)} - \lambda[I]^{(s)})\,\{\eta\}^{(s)} = {}^t\left[Y_j\right]^{(s)}\{f_j\}^{(s)} \qquad [3.49]$$

The equation of motion corresponding to the junction d.o.f.:

$$\{\mathrm{y_j}\}^{(s)} = \left[\mathrm{Y_j}\right]^{(s)}\{\eta\}^{(s)} + \left[\mathrm{R_{jj}}\right]^{(s)}\{\mathrm{f_j}\}^{(s)} \qquad [3.50]$$

By substituting the two equations [3.50] for the two substructures in equation [3.49], we obtain:

$$\{\mathrm{f_j}\}^{(1)} = [\mathrm{R}_{jj}^{(1)} + \mathrm{R}_{jj}^{(2)}]^{-1}\,[\mathrm{Y}_j^{(2)}\{\eta\}^{(2)} - \mathrm{Y}_j^{(1)}\{\eta\}^{(1)}] \qquad [3.51]$$

From equation [3.47] and from the two equations [3.49], the following equations can be derived:

$$[-\lambda \mathrm{I}^{(1)} + \Omega^{(1)}]\{\eta\}^{(1)} = -{}^t\mathrm{Y}_j^{(1)}[\mathrm{R}_{jj}^{(1)} + \mathrm{R}_{jj}^{(2)}]^{-1}\,[\mathrm{Y}_j^{(2)}\{\eta\}^{(2)} - \mathrm{Y}_j^{(1)}\{\eta\}^{(1)}] \qquad [3.52a]$$

$$[-\lambda \mathrm{I}^{(2)} + \Omega^{(2)}]\{\eta\}^{(2)} = -{}^t\mathrm{Y}_j^{(2)}[\mathrm{R}_{jj}^{(1)} + \mathrm{R}_{jj}^{(2)}]^{-1}\,[\mathrm{Y}_j^{(1)}\{\eta\}^{(1)} - \mathrm{Y}_j^{(2)}\{\eta\}^{(2)}] \qquad [3.52b]$$

Hence, the eigenvalue problem for the study of the coupled system:

$$\begin{bmatrix} \bar{K}_{11} & \bar{K}_{12} \\ \bar{K}_{21} & \bar{K}_{22} \end{bmatrix} \begin{bmatrix} \eta^{(1)} \\ \eta^{(2)} \end{bmatrix} = \lambda \begin{bmatrix} I^{(1)} & 0 \\ 0 & I^{(2)} \end{bmatrix} \begin{bmatrix} \eta^{(1)} \\ \eta^{(2)} \end{bmatrix} \quad [3.53a]$$

with:

$$\bar{K}_{11} = \Omega^{(1)} + {}^{t}Y_j^{(1)} \left[k_{jj}\right] Y_j^{(1)} \quad \bar{K}_{22} = \Omega^{(2)} + {}^{t}Y_j^{(2)} \left[k_{jj}\right] Y_j^{(2)} \quad [3.53b]$$

$$\bar{K}_{12} = {}^{t}\bar{K}_{21} = -\,{}^{t}Y_j^{(1)} \left[k_{jj}\right] Y_j^{(2)} \quad \left[k_{jj}\right] = [R_{jj}^{(1)} + R_{jj}^{(2)}]^{-1}$$

This type of method has advantages compared to the fixed-interface method. The resolution of the condensed problem indeed gives a higher accuracy with a rather better numerical efficiency (solving speed). In addition, it is compatible with analytical and experimental data.

However, the final condensed problem of this method is a function of the generalized coordinates alone. On the one hand, this property is detrimental when we are interested in a few physical d.o.f. imposed in the condensed model, and, on the other hand, when willing to compute forced responses. It is actually necessary to shift back from generalized coordinates to physical d.o.f. The other major issue arises when assembling more than two substructures. Indeed, the condensed problem is not of the $\left[L_c\right] = {}^{t}[Z][L][Z]$ ([L] = [K], [M]) form, thereby creating problems when using this method in model registration and optimization procedures. In this case, the assembly techniques presented above can no longer be applied and consequently, only manual procedures, which can prove time-consuming and costly, are considered for the explicit calculation of assembled condensed matrices. In the following, a method of condensation by dynamic multiplication can be used using the free-interface modes of the substructures [TRA 01]. This technique has the advantage over Mac Neal's method of preserving the junction d.o.f., enabling the substructures to be assembled in a simple and systematic way.

3.3.5. *Open interface method*

In this case, the motion of each substructure can be characterized by its free-interface modes, its rigid body modes when they exist and its attachment modes

[TRA 01]. Hence, the basic relation of each substructure, the index(s) has been omitted from equations [3.54]–[3.68]:

$$\{y\} = [Y]\{\eta\} + [\psi_r]\{\xi_r\} + [\psi_a]\{\xi_a\} \quad [3.54]$$

[Y] is the truncated modal basis of the free-interface substructure, $[\psi_r]$ is the matrix of rigid body modes and $[\psi_a]$ is the matrix of attachment modes obtained by computing the displacements produced by unit forces applied to the junction d.o.f. assuming that the other junction d.o.f. is fixed obtained by the relation:

$$[\psi_a] = [G][F_j] \quad [3.55a]$$

with:

$$[F_j] = \begin{bmatrix} 0 \\ I_j \end{bmatrix} \quad [3.55b]$$

The expression of matrix [G] depends on the nature of the structure [BRA 96], in the case of a structure that does not contain rigid field modes:

$$[G] = [K]^{-1} \quad [3.56]$$

In the presence of a structure with rigid field modes, this matrix becomes:

$$[G] = {}^t[A][K_c]^{-1}[A] \quad [3.57a]$$

with:

$$[A] = [I] - [\varphi^{(r)}]^t [\varphi^{(r)}][M] \quad [3.57b]$$

$${}^t[\psi_r][M][\psi_r] = [I] \quad [3.57c]$$

$[K_c]$ is the stiffness matrix obtained by fixing arbitrary d.o.f. to make the structure isostatic. That is, by replacing the corresponding part of the stiffness matrix of the initial structure with zero. To keep the junction d.o.f., we use the following decomposition:

$$[Y] = \begin{bmatrix} Y_i \\ Y_j \end{bmatrix} \quad [\psi_r] = \begin{bmatrix} \psi_{ri} \\ \psi_{rj} \end{bmatrix} \quad [\psi_a] = \begin{bmatrix} \psi_{ai} \\ \psi_{aj} \end{bmatrix} \qquad [3.58]$$

Equation [3.54] can be rewritten as:

$$\begin{bmatrix} y_i \\ y_j \end{bmatrix} = \begin{bmatrix} Y_i \\ Y_j \end{bmatrix} \{\eta\} + \begin{bmatrix} \psi_{ri} \\ \psi_{rj} \end{bmatrix} \{\xi_r\} + \begin{bmatrix} \psi_{ai} \\ \psi_{aj} \end{bmatrix} \{\xi_a\} \qquad [3.59]$$

Expanding the equations for the first group of rows yields:

$$\{\xi_a\} = \psi_{aj}^{-1}\{y_j\} - \psi_{aj}^{-1} Y_j \{\eta\} + \psi_{aj}^{-1} \psi_{rj} \{\xi_r\} \qquad [3.60]$$

Substituting $\{\xi_a\}$ in equation [3.54] leads to:

$$\{y\} = \psi_a \psi_{aj}^{-1}\{y_j\} + (\psi_r - \psi_a \psi_{aj}^{-1} \psi_{rj})\{\xi_r\} + (Y - \psi_a \psi_{aj}^{-1} Y_j)\{\eta\} \qquad [3.61]$$

The matrix [Q] then becomes in this case:

$$[Q] = \begin{bmatrix} \psi_a \psi_{aj}^{-1} & \psi_r - \psi_a \psi_{aj}^{-1} \psi_{rj} & Y - \psi_a \psi_{aj}^{-1} Y_j \end{bmatrix} \qquad [3.62]$$

The residual attachment modes are obtained by removing attachment modes from the free-interface modes already kept in the truncated modal basis of the free-interface substructure, in this case:

$$\{y\} = [Y]\{\eta\} + [\psi_r]\{\xi_r\} + [\psi_{ar}]\{\xi_{ar}\} \qquad [3.63]$$

$[\psi_{ar}]$ is the matrix of residual attachment modes obtained by:

$$[\psi_{ar}] = [R][F_j] \qquad [3.64a]$$

with:

$$[R] = [G] - [Y][\Omega]^{-1}\,{}^t[Y] \qquad [3.64b]$$

where $[\Omega]$ is the matrix of the eigenvalues kept.

The matrix [Q] then becomes:

$$[\mathrm{Q}] = \begin{bmatrix} \psi_{ar}\psi_{arj}^{-1} & \psi_r - \psi_{ar}\psi_{arj}^{-1}\psi_{rj} & \mathrm{Y} - \psi_{ar}\psi_{arj}^{-1}\mathrm{Y}_j \end{bmatrix} \qquad [3.65]$$

3.3.6. *Hybrid method*

In this case, the interface is divided into a fixed interface and a free interface [TRA 01]. The motion of each substructure can be characterized by its eigenmodes, its static joint modes and its attachment modes. Hence, the basic relation for each substructure is given by:

$$\{y\} = [\mathrm{Y}]\{\eta\} + \begin{bmatrix}\psi_{cj}\end{bmatrix}\{y_{jf}\} + \begin{bmatrix}\psi_{aj}\end{bmatrix}\{\xi_{jl}\} \qquad [3.66]$$

where [Y] is the truncated modal basis of the substructure $SS^{(s)}$; $\begin{bmatrix}\psi_{cj}\end{bmatrix}$ is the matrix of static joint modes obtained by calculating the displacements corresponding to a unit displacement in a fixed-interface d.o.f. assuming that the free-interface d.o.f. are free and $\begin{bmatrix}\psi_{aj}\end{bmatrix}$ is the attachment mode matrix obtained by calculating the displacements produced by unit forces applied to the free-interface d.o.f. assuming that the interface d.o.f. are fixed.

In this case, the matrix [Q] becomes:

$$[\mathrm{Q}] = \begin{bmatrix} \psi_a\psi_{aj}^{-1} & \psi_{cj} - \psi_a\psi_{aj}^{-1}\psi_{cj} & \mathrm{Y} - \psi_a\psi_{aj}^{-1}\mathrm{Y}_j \end{bmatrix} \qquad [3.67]$$

If we use the residual attachment modes, this matrix becomes:

$$[\mathrm{Q}] = \begin{bmatrix} \psi_{ar}\psi_{arj}^{-1} & \psi_c - \psi_{ar}\psi_{arj}^{-1}\psi_{cj} & \mathrm{Y} - \psi_{ar}\psi_{arj}^{-1}\mathrm{Y}_j \end{bmatrix} \qquad [3.68]$$

3.3.7. *Reduction of junction d.o.f.*

One of the problems with fixed-interface or free-interface multiplication methods is that all of the junction d.o.f. between the substructures are preserved. This significantly increases the size of the condensed model. Dynamic multiplication methods remain sometimes inadequate to reduce the order of the final condensed system, and the condensation of junction d.o.f. becomes necessary [BRA 96; BOU 92; TRA 01].

In this section, a method for reducing junction d.o.f. is proposed based on the use of interface modes. These modes are obtained from the condensation at the Guyan interfaces of the full structure.

The vector of the displacement of the overall structure is as follows:

$\{y_j\}$ (jx1) is the subvector of the master junction d.o.f. to be kept;

$\{y_i\}$ (ix1) is the subvector of the internal slave d.o.f. to be eliminated.

The modal equation of motion of the full system becomes:

$$\left(\begin{bmatrix} K_{ii} & K_{ij} \\ {}^tK_{ij} & K_{jj} \end{bmatrix} - \lambda \begin{bmatrix} M_{ii} & M_{ij} \\ {}^tM_{ij} & M_{jj} \end{bmatrix}\right)\begin{Bmatrix} y_i \\ y_j \end{Bmatrix} = \begin{Bmatrix} 0 \\ f_j \end{Bmatrix} \quad [3.69]$$

It is assumed that there are no external forces on the internal d.o.f. Expanding the equations for the first group of rows, a relationship between $\{y_j\}$ and $\{y_i\}$ is obtained such as:

$$\{y_i\} = \left[t_{ij}(\lambda)\right]\{y_j\} \quad [3.70a]$$

with:

$$\left[t_{ij}(\lambda)\right] = -(K_{ii} - \lambda M_{ii})^{-1}(K_{ij} - \lambda M_{ij}) \quad [3.70b]$$

The method proposed by Guyan [GUY 65] is widely used in the dynamic analysis of structures, and only uses the following static transformation ($\lambda = 0$):

$$\left[t_{ij}\right] = -K_{ii}^{-1}K_{ij} \quad [3.71]$$

The transformation equation is expressed only in terms of junction d.o.f. (masters):

$$\{y\} = \begin{Bmatrix} y_i \\ y_j \end{Bmatrix} = \begin{bmatrix} t_{ij} \\ I_j \end{bmatrix}\{y_j\} = \left[T_{ij}\right]\{y_j\} \quad [3.72]$$

This transformation applied to the initial problem of the global structure results in obtaining a condensed problem such as:

$$\left(\left[\mathrm{K}_{cj}\right]-\lambda\left[\mathrm{M}_{cj}\right]\right)\left\{\mathrm{y}_{j}\right\}=0 \qquad [3.73a]$$

with:

$$\left[\mathrm{K}_{cj}\right]={}^{t}\left[\mathrm{T}_{ij}\right][\mathrm{K}]\left[\mathrm{T}_{ij}\right] \; \left[\mathrm{M}_{cj}\right]={}^{t}\left[\mathrm{T}_{ij}\right][\mathrm{M}]\left[\mathrm{T}_{ij}\right] \qquad [3.73b]$$

The displacement vector of the interface can be expressed by:

$$\left[\mathrm{y}_{j}\right]=\left[\varphi_{\mathrm{j}}\right]\left\{\eta_{\mathrm{j}}\right\} \qquad [3.74]$$

where $\left[\varphi_{\mathrm{j}}\right]$ is the truncated matrix of the eigenmodes of the interface.

The condensed vector of the displacements of the assembled structure becomes:

$$\left\{\mathrm{y}_{c}\right\}=\begin{Bmatrix}\eta_p^{(1)}\\ \vdots\\ \eta_p^{(K)}\\ \mathrm{y}_j\end{Bmatrix}=\begin{bmatrix}I^{(1)} & \cdots & 0 & 0\\ \vdots & \ddots & & \\ 0 & \cdots & I^{(K)} & \\ 0 & \cdots & & \varphi_j\end{bmatrix}\begin{Bmatrix}\eta_p^{(1)}\\ \vdots\\ \eta_p^{(K)}\\ \eta_j\end{Bmatrix} \qquad [3.75]$$

In this case, the transformation matrix for each substructure becomes:

$$[\mathrm{Z}]^{(k)}=[\mathrm{Q}]^{(k)}[\beta]^{(k)}[\mathrm{T}] \qquad [3.76a]$$

with:

$$[T]=\begin{bmatrix}I^{(1)} & \cdots & 0 & 0\\ \vdots & \ddots & & \\ 0 & \cdots & I^{(K)} & \\ 0 & \cdots & & \varphi_j\end{bmatrix} \qquad [3.76b]$$

3.4. Stochastic finite element methods

3.4.1. *Introduction*

When considering the study of mechanical systems by the finite element method, one of the main assumptions is that the model is deterministic. As the performance of computing machines considerably increases over time, it becomes possible to take into account complex phenomena that were previously overlooked or simplified. In particular, it becomes possible to work with problems for which some data are uncertain. This allows the intrinsic uncertainties of the properties of materials and structures to be taken into account. The need to take these uncertainties into account leads to the development of stochastic finite element methods and reliability-mechanical couplings.

Two main families of numerical methods can currently be found:

– The methods of the first two moments permit a sensitivity analysis of the structural response whose parameters are uncertain by combining the finite element method and the theory of probabilities.

– Reliability-mechanical couplings enable a reliability analysis of a complex structure by combining the finite element method and reliability methods.

Regardless of the method of analysis used, the uncertainties in the mechanical properties of the materials, the geometric characteristics of the structure and the applied loading vary according to their spatial position. These uncertainties are then modeled by random fields. However, reliability and sensitivity analysis requires a discretization of random fields [FLO 94].

3.4.2. *Discretization of random fields*

This process transforms the different fields into vectors of correlated random variables. Let v(x) be a multidimensional random field defined in a domain Ω. We consider that the field is completely described by its average E(x), by its variance $\sigma^2(x)$ and by its correlation function $\rho(x,x')$ between two distinct points of coordinates x and x'. The discretization of the field v(x) consists of replacing it with a random vector v. All existing methods relate v(x) to v by means of a linear transformation. Therefore, for a Gaussian random field, v is also Gaussian. There are many methods for the discretization of random fields [FLO 94]. These methods use a stochastic mesh of the domain Ω. This is not necessarily identical to the finite element mesh of the structure.

3.4.2.1. *Method of discretization of median points*

The simplest method of discretization is that of the median point [YAM 88]. The value of the field within the domain Ω_e for an element is described by a simple random variable that represents the value of the field at the central point of the element, that is, at the center x_c.

Then:

$$\tilde{v}(x) = v(x_c) \qquad x \in \Omega_e$$

The random variables v(xc) for all elements form the vector v. In general, this discretization method tends to overrepresent the variability within an element.

3.4.2.2. *Spatial averaging discretization method*

This method describes the field inside each element according to the spatial average of the field on the element [NPV 83], that is,

$$\tilde{v}(x) = \frac{\int_{\Omega_e} v(x) d\Omega_e}{\int_{\Omega_e} d\Omega_e} \qquad x \in \Omega_e \qquad [3.78]$$

The set of random variables $v(x_c)$ for all elements forms the vector v. In general, the variance of the spatial mean is greater than that of the median point. This method tends to underrepresent the variability within an element.

3.4.2.3. *Discretization method by shape function*

This method describes the random field in an element based on a set of nodal values and its corresponding shape function [KLE 92].

Then:

$$\tilde{v}(x) = \sum_{i}^{n} N_i(x) v(x_i) \qquad x \in \Omega_e \qquad [3.79]$$

n is the number of nodes of the element, x_i represents the coordinates of node i, and Ni(x) is the i-th shape function, usually a polynomial. The nodal values for the entire field form the vector v in this case. The realization of the discretized field in this method is a continuous function that represents an advantage over the previous methods.

3.4.2.4. *Optimal linear estimation discretization method*

The random field is described by a linear function of the nodal values $v(x_i)$ of the form [DER 91]:

$$\tilde{v}(x) = a(x) + \sum_{i}^{n} {}^{T}b_i(x)v(x_i) \qquad x \in \Omega_e \qquad [3.80]$$

n is the number of nodes in the element. The functions a(x) and $b_i(x)$ are optimally determined by minimizing the variance of the error $\tilde{v}(x) - v(x)$, $\tilde{v}(x)$ being an unbiased estimator of the mean. This discretization method is currently one of the most efficient.

3.4.3. *Methods for calculating moments*

The goal of the methods of second-order moments is to calculate the mean and variance of the structural response [CAM 00]. All of these methods have the particularity of being limited to the moments of the variables and do not reflect the influence of the type of distributions of the variables. Methods for calculating moments therefore provide, by nature, limited information on the variability of the response function.

3.4.3.1. *Monte Carlo simulation*

The estimation of the moments of the response function F can be obtained by Monte Carlo simulation [SHI 72]. F is seen as the random variable image of the variables of basis x_i. The simulations consist of constructing a sample (F_1, F_2,..., F_n) of the random variable F, and processing this sample using the usual statistical techniques. The n simulations are performed independently according to the distribution of the random vector x.

The estimator of the mean E(F) of F is:

$$E(F) = \frac{1}{n}\sum_{i=1}^{n} F_i \qquad [3.81]$$

The variance estimator var(F) of F is:

$$\operatorname{var}(F) = \frac{1}{n-1}\sum_{i=1}^{n} \left[F_i - E(F)\right]^2 \qquad [3.82]$$

Simulation methods require all the more simulations for larger response coefficient of variation. The precision of the results is independent of the number of variables.

3.4.3.2. *Methods of perturbation*

The perturbation method is widely used in the field of stochastic finite elements. Based on a Taylor series expansion of the structure response with respect to the basic physical random variables, mechanical properties, geometric features or applied forces. Perturbation methods calculate the mean and variance (covariance matrix) of the response of a mechanical structure with uncertain variables. This method is used in many fields, to solve both static and dynamic linear and nonlinear problems [LIU 86]. The calculation of the first two statistical moments by a Taylor expansion of the frequency response is presented in [SIN 93]. Unlike the system response, modal quantities (for instance, frequencies and eigenmodes) are slowly variable according to the properties of the system. Perturbation-based approaches with these quantities have therefore been experimented with [KLE 92]. In [MUS 99], a clever first-order perturbation method seems to obtain results about as good as by a second-order technique but with much less computations for a large number of parameters (since they are first-order ones). Taylor's first-order series expansion appears to be reasonably adequate for weak basic random variables. It is efficient in terms of computation time. This method has a serious drawback of being limited to problems whose parameters are slightly variable and whose nonlinearities with respect to the parameters are too strong. The greater the nonlinearity of the response function with respect to these variables and the greater the variability of these parameters, the greater the order of Taylor series expansion must be.

Therefore, according to Dr. Shinozuka, second-order perturbation methods require more computation than a variability study using the Monte Carlo method [SHI 88]. As a matter of fact, the calculation of the second-order derivatives of the response function with respect to the basic random variables is particularly cumbersome to perform. Moreover, second-order terms affect only the mean values of the response and are negligible with zeroth- and first-order terms [TEI 91]. A Taylor series expansion of the response function with respect to the basic random variables requires the computation of the first derivatives, at the first order, or second derivatives, at the second order of the response function. The accuracy of the results, mean and variance in an analysis of the first two moments thus strongly depends on the calculation of these gradients. In the following, we present two perturbation methods developed to solve linear static problems. The first is based on a Taylor series expansion [FLO 94], and the second is based on a clever first-order expansion [MUS 99].

a) Taylor's series expansion

The structural response is the vector of the nodal displacements.

Let the equilibrium equation of a structure with linear behavior be:

$$[\mathrm{K}]\{\mathrm{y}\} = \{\mathrm{F}\} \qquad [3.83]$$

where $\{y\}$ is the vector of nodal displacements of dimensions (Nx1). [K] is the stiffness matrix of the structure (NxN). $\{F\}$ is the vector of the equivalent nodal external forces of dimensions (Nx1). For a structure with uncertain variables, it is assumed that the stiffness matrices [K] as well as the loading vector $\{F\}$ are functions of the vectors of the random variables α_p (p=1,..., P).

We define the vector of the mean parameters $\{\bar{\alpha}\}$, and the quantity $\{d\alpha\} = \{\alpha\} - \{\bar{\alpha}\}$. To simplify the following notation is adopted for the derivatives of a quantity A:

$$[A]^0 = [A](\{\bar{\alpha}\}) \quad [\mathrm{A}]^n = \left.\frac{\partial[\mathrm{A}]}{\partial\alpha_n}\right|_{\{\bar{\alpha}\}} \quad [\mathrm{A}]^{np} = \left.\frac{\partial^2[\mathrm{A}]}{\partial\alpha_n\partial\alpha_p}\right|_{\{\bar{\alpha}\}} \qquad [3.84]$$

The expansion of matrix [K], the vector of the external nodal forces $\{F\}$ and the vector of the nodal displacements in second-order Taylor series with respect to the random variable P $d\alpha_p$ in the neighborhood of the vector of the mean parameters $\{\bar{\alpha}\}$:

$$[\mathrm{K}] = [\mathrm{K}]^0 + [\mathrm{K}]^n\, d\alpha_n + \frac{1}{2}[\mathrm{K}]^{np}\, d\alpha_n d\alpha_p \qquad [3.86a]$$

$$\{\mathrm{F}\} = \{\mathrm{F}\}^0 + \{\mathrm{F}\}^n\, d\alpha_n + \frac{1}{2}\{\mathrm{F}\}^{np}\, d\alpha_n d\alpha_p \qquad [3.86b]$$

$$\{\mathrm{y}\} = \{\mathrm{y}\}^0 + \{\mathrm{y}\}^n\, d\alpha_n + \frac{1}{2}\{\mathrm{y}\}^{np}\, d\alpha_n d\alpha_p \qquad [3.86c]$$

where $[\ ]^0$, $[\ ]^n$ and $[\ ]^{np}$ are the deterministic matrices.

We should note that for the index notation that the repetition of the indices n, p twice implies a summation. By injecting equations [3.85] into the equilibrium equation [3.84] and grouping the terms of the same order, we obtain:

$$\{y\}^0 = [K]^{0-1}\{F\}^0 \quad [3.86a]$$

$$\{y\}^n = [K]^{0\text{-}1}(\{F\}^n - [K]^n\{y\}^0) \quad [3.86b]$$

$$\{y\}^{np} = [K]^{0\text{-}1}(\{F\}^{np} - [K]^n\{y\}^p - [K]^p\{y_n\} - [K]^{np}\{y\}^0) \quad [3.86c]$$

The calculation of these three vectors therefore requires only one evaluation of the inverse of the stiffness matrix $[K]^{0\text{-}1}$, hence a significant gain in computational time. The mean value or the expected value of the vector of the nodal displacements is then:

$$E[\{y\}] = \{y\}^0 + \frac{1}{2}\{y\}^{np}\operatorname{cov}(\alpha_n, \alpha_p) \quad [3.87a]$$

and its covariance matrix:

$$\operatorname{cov}(\{y_i\},\{y_j\}) = \{y_i\}^n\{y_j\}^p\operatorname{cov}(\alpha_n, \alpha_p) \quad [3.87b]$$

b) Muscolino perturbation method

In this method, the random parameters are assumed to be uncorrelated [MUS 99]. For a quantity [A], the following notations will be used:

$$\left[\bar{A}\right] = [A]\Big|_{\{\bar{\alpha}\}} + \frac{1}{2}E[\alpha_n^2]\left.\frac{\partial^2[A]}{\partial\alpha_n^2}\right|_{\{\bar{\alpha}\}} \qquad [A]^n = \left.\frac{\partial[A]}{\partial\alpha_n}\right|_{\{\bar{\alpha}\}} \quad [3.88]$$

The development of the different quantities according to the Muscolino method:

$$[K] = \left[\bar{K}\right] + [K]^n d\alpha_n \quad [3.89a]$$

$$\{y\} = \{\bar{y}\} + \{y\}^n d\alpha_n$$

$$\{\mathrm{F}\} = \{\overline{\mathrm{F}}\} + \{F\}^{n}\, d\alpha_n \qquad [3.89b]$$

We replace equation [3.89] in equation [3.83], when the terms of the same order are grouped together. The following equations are obtained, if the inverse of $\left[\overline{\mathrm{K}}\right]$ exists:

$$\{\overline{\mathrm{y}}\} = \left[\overline{\mathrm{K}}\right]^{-1}\{\overline{\mathrm{F}}\} \qquad [3.90a]$$

$$\{\mathrm{y}\}^{n} = \left[\overline{\mathrm{K}}\right]^{-1}(\{\mathrm{F}\}^{n} - [\mathrm{K}]^{n}\{\overline{\mathrm{y}}\}) \qquad [3.90b]$$

The mean value or the expected value of the vector of the nodal displacements is then:

$$E[\{\mathrm{y}\}] = \{\overline{\mathrm{y}}\} \qquad [3.91a]$$

and its covariance matrix is written as:

$$\mathrm{cov}(\{y_i\},\{\mathrm{y_j}\}) = \{y_i\}^{n}\{\mathrm{y}_j\}^{n}\,\mathrm{var}(\alpha_n) \qquad [3.91b]$$

3.4.3.3. *Neumann's method*

To overcome the perturbation problems, some authors have proposed to couple a Neumann expansion method to Monte Carlo simulations [FLO 94]. The stiffness matrix $[\mathrm{K}]$ is decomposed into:

$$[\mathrm{K}] = [\mathrm{K}_0] + \Delta[\mathrm{K}] \qquad [3.92]$$

The coefficients of the matrix $[\mathrm{K}_0]$ are constant in which the random parameters are replaced by their means, the coefficients of $\Delta[\mathrm{K}]$ gather the variations of these same parameters that can be obtained by a direct Monte Carlo simulation (SMC).

Similarly, the force vector $\{\mathrm{F}\}$ can be decomposed as:

$$\{\mathrm{F}\} = \{F_0\} + \Delta\{\mathrm{F}\} \qquad [3.93]$$

The vector $\{y_0\}$ is obtained by solving:

$$[K_0]\{y_0\} = \{F_0\} + \Delta\{F\} \quad [3.94]$$

The Neumann expansion of the inverse stiffness matrix (if it exists) is written as:

$$\begin{aligned}[K]^{-1} &= ([K_0] + \Delta[K])^{-1} \\ &= ([I] - [P] + [P]^2 - [P]^3 + \ldots\ldots)[K_0]^{-1}\end{aligned} \quad [3.95a]$$

where:

$$[P] = [K_0]^{-1}\Delta[K] \quad [3.95b]$$

Substituting $[K]^{-1}$ in equation [3.83] leads to:

$$\{y\} = (I - [P] + [P]^2 - [P]^3 + \ldots\ldots)[K]^{-1}(\{F_0\} + \Delta\{F\}) \quad [3.96]$$

We define:

$$\{\bar{y}_0\} = [K_0]^{-1}\{F_0\} \;\; \Delta\{y_0\} = [K_0]^{-1}\Delta\{F_0\} \quad [3.97]$$

Equation [3.96] becomes:

$$\begin{aligned}\{y\} &= \{\bar{y}_0\} - [P]\{\bar{y}_0\} + [P]^2\{\bar{y}_0\} - [P]^3\{\bar{y}_0\} + \ldots\ldots. \\ &\quad + \Delta\{y_0\} - [P]\Delta\{y_0\} + [P]^2\Delta\{y_0\} - [P]^3\Delta\{y_0\} + \ldots.. \\ &= \sum_{r=0}^{\infty}(-1)^r[P]^r(\{\bar{y}_0\} + \Delta\{y_0\})\end{aligned} \quad [3.98]$$

Using the following equalities:

$$\{\bar{y}_r\} = [P]^r\{\bar{y}_0\} \;\; \Delta\{y_r\} = [P]^r\Delta\{y_0\} \; \{y_r\} = \{\bar{y}_r\} + \Delta\{y_r\} \quad [3.99]$$

The random vector $\{y\}$ can be written as:

$$\begin{aligned}\{y\} &= \sum_{r=0}^{\infty} (-1)^r \{y_r\} \\ &= \{y_0\} - \{y_1\} + \{y_2\} - \{y_3\} + \ldots\ldots\end{aligned} \quad [3.100]$$

and the vector $\{y_r\}$ is given by:

$$\{y_r\} = [P]^r (\{\bar{y}_0\} + \Delta\{y_0\}) = [P]([P]^{r-1}\{y_0\}) = [P]\{y_{r-1}\} \quad r=1,2,3,\ldots. \quad [3.101]$$

The vector $\{y\}$ is entirely determined from equation [3.94] which enables us to calculate$\{y_0\}$, and the recursion process [3.101]. A convergence test ensures that the method is stopped

$$\frac{\left\|\{y_r\}\right\|_2}{\left\|\sum_{k=0}^{r} (-1)^k \{y_k\}\right\|_2} \leq \delta_{err} \quad [3.102]$$

where δ_{err} represents the admissible error for convergence.

Using the Monte Carlo simulation to calculate $y_r^{(i)}$ (i=1.2,....,Ns; r=1,2,3,.....) in equation [3.101] and we deduce $y^{(i)}$ from equation [3.102]. The mean value of$\{y\}$ is given by:

$$E[\{y\}] = \{\bar{y}\} = \frac{1}{N_s} \sum_{i=1}^{N_s} \{y\}^{(i)} \quad [3.103a]$$

The variance of $\{y\}$ is given by:

$$Var[\{y\}] = \frac{1}{N_s} \sum_{i=1}^{N_s} diag[(\{y\}^{(i)} - \{\bar{y}\}) \times {}^t(\{y\}^{(i)} - \{\bar{y}\})] \quad [3.103b]$$

We should note that this approach converges only if the absolute values of all characteristic values of $[P] = [K_0]^{-1} \Delta[K]$ are less than 1.

3.4.3.4. *Projection on polynomial chaos*

Stochastic solutions are projected on a basis of orthogonal polynomials whose variables are orthonormal Gaussians [DES 00]. The properties of this polynomial basis are used to generate a system of linear equations by means of projection. The solution to this system leads to an expansion of the solution on the polynomial basis, which can be used to analytically or numerically find either the mean, the standard deviation or the distribution of the random solution.

The chaotic polynomials ψ_m corresponding to the multidimensional Hermite polynomials obtained by:

$$\psi_m(\xi_1,\ldots,\xi_n) = (-1)^n \exp(\frac{1}{2}{}^t\{\xi\}\{\xi\})\frac{\partial^n}{\partial\xi_1\cdots\xi_n}\exp(-\frac{1}{2}{}^t\{\xi\}\{\xi\}) \qquad [3.104]$$

The method used is now developed to find the expansion to homogeneous chaos of the solution of a linear static problem of which certain parameters are random. The random stiffness matrix [K] can be written as:

$$\left[K\right]=\left[K_0\right]+\left[\tilde{\mathrm{K}}\right] \qquad [3.105]$$

The matrix $\left[K_0\right]$ is a deterministic matrix. The matrix $\left[\tilde{\mathrm{K}}\right]$ corresponds to the random part of the stiffness matrix $\left[\tilde{\mathrm{K}}\right]$ and is rewritten from a Karhunen–Loeve expansion [GHA 91] as:

$$\left[\tilde{\mathrm{K}}\right]=\sum_{q=1}^{Q}\left[K_q\right]\xi_q \qquad [3.106]$$

where the ξ_q are independently reduced centered Gaussians, which can be made to correspond to the first polynomials Ψ_q, and the matrices [K_q] are the deterministic matrices. We set $\xi_0 = 1$. We can write the random matrix [K] from equations [3.106] and [3.107] as:

$$\left[\mathrm{K}\right]=\sum_{q=0}^{Q}\left[K_q\right]\xi_q \qquad [3.107]$$

The vector of the applied random forces can be written as:

$$\{F\} = \{F_0\} + \{\tilde{F}\} \qquad [3.108]$$

where $\{F_0\}$ is the deterministic force vector and $\{\tilde{F}\}$ is the random force vector.

From a Karhunen–Loeve expansion, the random force vector is written as:

$$\{\tilde{F}\} = \sum_{p=1}^{P} \{F_p\}\xi_p \qquad [3.109]$$

which can be rewritten as:

$$\{F\} = \sum_{p=0}^{P} \{F_p\}\xi_p \qquad [3.110]$$

in which $\xi_0 = 1$. A decomposition of the displacement vector $\{y\}$ on polynomials with Q orthonormal Gaussian random variables is:

$$\{y\} = \sum_{n=0}^{N} \{y_n\}\Psi_n(\{\xi_i\}_{i=1}^{Q}) \qquad [3.111]$$

We then replace $\{y\}$ by its expansion in equation [3.83], and we multiply the equation obtained by Ψ_m. If we average (integration on the domain of random variables) using the orthogonality properties of the polynomials, it follows that:

$$\sum_{q=0}^{Q} \sum_{n=0}^{N} \{y_n\}\langle \xi_q \ \Psi_n \ \Psi_m \rangle [K_q] = \sum_{p=0}^{P} \{F_p\}\langle \Psi_m \ \xi_p \rangle \qquad [3.112]$$

System [3.112] can be written as:

$$\begin{bmatrix} D^{(00)} & \cdots & D^{(0N)} \\ . & \cdots & . \\ . & D^{(jk)} & . \\ . & \cdots & . \\ D^{(N0)} & \cdots & D^{(NN)} \end{bmatrix} \begin{Bmatrix} y_0 \\ . \\ y_k \\ . \\ y_N \end{Bmatrix} = \begin{Bmatrix} f_0 \\ . \\ f_k \\ \vdots \\ f_N \end{Bmatrix} \qquad [3.113a]$$

in which, we have:

$$[K]^{(jk)} = \sum_{q=0}^{Q} \langle \xi_q \psi_j \psi_k \rangle [K_q] \tag{3.113b}$$

$$\{f_k\} = \sum_{p=0}^{P} \langle \xi_p \psi_k \rangle \{F_p\} \tag{3.113c}$$

It should be noted that due to the orthogonality of polynomials, most expressions $\langle \xi_q \Psi_n \Psi_m \rangle$ are zero. Once the $\{y_n\}$ are known, the average is directly given by:

$$E[\{y\}] = \{y_0\} \tag{3.114a}$$

and the variance is equal to:

$$Var[\{y\}] = \sum_{n=1}^{N} (\{y_n\})^2 \langle \Psi_n^2 \rangle \tag{3.114b}$$

If the statistical distribution of the possible values of $\{y\}$ is also of interest, a Monte Carlo simulation can be carried out. This type of simulation is usually avoided for computational cost reasons, but for a projection on homogeneous chaos, only a few terms can be simulated, instead of having to invert an entire system each time it is sampled. The amount of calculation is therefore quite reasonable, since for each sampling it consists of calculating the values of the polynomials Ψ_n multiplied by $\{y_n\}$. It is therefore possible to consider using a large sample to obtain the interesting statistical values of the solution. For example, the mean, the standard deviation, the distribution, the most likely value and a histogram of the values of $\{y\}$.

3.5. Conclusion

In this chapter, we have presented different stochastic finite element methods and modal synthesis methods. Nonetheless, it should be noted that classical substructuring methods have a major disadvantage which lies in the preservation of the junction d.o.f. in the state vector of the final condensed system. A method for reducing these junction d.o.f. has been proposed. These modes are obtained from the condensation at the Guyan interfaces of the full structure.

4

Fatigue and Reliability Optimization for Structures Subjected to Random Vibrations

4.1. Introduction

Determining the reliable design of a mechanical structure is one of the major concerns for an engineer. The objective is to find the best design that leads to reducing costs and satisfying system performance.

The increase in the performance of super-computers for complex structures has allowed the development of numerical optimization techniques. Generally, the optimization of mechanical structures can be mathematically modeled, by a problem of the minimization of an objective function under constraints, which are named limit-state functions. This problem is generally solved with methods that make use of iterative algorithms [FLE 87].

In the state of the art, structural optimization techniques can be divided into three main families:

– Design optimization: the structural model with a fixed shape and topology can be improved. The design variables are therefore the dimensions of the components forming the structure, such as beam lengths and cross-sections or a plate thickness.

– Shape optimization: it makes it possible to modify the geometry of the structure. The topology is set at the beginning. Design variables are therefore parameters representing the boundaries of the domain.

– Topology optimization: this allows controlling the topology and geometry of the part. It is often referred to as generalized form optimization.

Optimizing structures without taking uncertainties into account can lead to failures or design states that do not meet certain reliability criteria. For this purpose, several methods of reliability optimization have been developed. The objective of such a method is therefore to design structures that should be both economical and reliable. These methods have been applied in different fields such as: biomechanics [ELH 15], civil engineering [KOU 20], mechatronics [MAK 16], vibro-acoustics [DAM 19] or aerodynamics [ELH 13], etc.

Many structures in industry are subject to random vibrations, and these vibrations can cause fatigue damage. In recent years, several methods have been developed to determine the lifetime of a structure subjected to random vibrations in order to minimize the computation time, damage analysis in the frequency domain, by the use of spectral methods. In structural dynamics, these methods require the calculation of the different statistical properties of random vibrations.

The objective is to determine a method for reliability optimization of structures subjected to random vibrations considering fatigue damage [ABI 19; YAI 18].

This section presents the methodology to be followed for the determination of fatigue damage of structures subjected to random vibrations. Next, the methodology for deterministic optimization is defined. This is followed by a presentation of the development of the different RBDO (Reliability Based Design Optimization) methods for the design of structures and a method called RHM (Robust Hybrid Method) developed in order to improve numerical performance compared to the classical method. An extension of this method to other fields of application is presented, and its performance is outlined.

4.2. Fatigue damage analysis

4.2.1. *Formulations and developments*

The primary purpose of multiaxial fatigue criteria is to determine critical points in a structure. Many authors may think that crack initiation is governed by the second invariant of the deflecting stress tensor [BEN 08]. This can clearly be seen in the Crossland criterion (equation [4.1]) and the Sines criterion (equation [4.2]) where $J_{2,a}$ is the second invariant of the stress deflector

$$D_{Crossland} = \frac{\sqrt{J_{2,a}} + \alpha \max\limits_{0 \le t \le T}\{p(t)\}}{\beta} \le 1 \qquad [4.1]$$

$$D_{Sines} = \frac{\sqrt{J_{2,a}} + \alpha\, \mathrm{E}[p(t)]}{\beta} \le 1 \qquad [4.2]$$

where p(t) is the hydrostatic pressure (equation [4.3]) and α, β are constants related to matter.

For the Sines criterion, they are defined by the alternating tensile endurance limit f_{-1}, the alternating torsional endurance limit t_{-1} and fracture strength R_m (equations [4.4] and [4.5])

$$\mathrm{p(t)} = \frac{1}{3}\left(S_{xx} + S_{yy} + S_{zz}\right) \quad [4.3]$$

$$\alpha = \frac{3t_{-1}(R_m + f_{-1})}{f_{-1}R_m} - \sqrt{6} \quad [4.4]$$

$$\beta = t_{-1} \quad [4.5]$$

The Crossland criterion requires the computation of the maximum hydrostatic pressure while the Sines criterion uses its average. This reduces the time it takes to compute the damage if it is calculated according to the Sines criterion and has almost the same result as the Crossland criterion [WEB 99].

In this section, a linear structure subjected to stationary ergodic Gaussian loads is considered. The crack will not begin before Ne the repetitions of the periodic random load of duration T if the average damage should not exceed the fatigue limit. The Sines criterion is then defined by:

$$E[D_{Sines}] = \frac{E\left[\sqrt{J_{2,a}}\right] + \alpha\, \mathrm{E[p(t)]}}{\beta} \leq 1 \quad [4.6]$$

The problem now is in computing the average of the shear stress $E\left[\sqrt{J_{2,a}}\right]$. In fact, for proportional loading, $\sqrt{J_{2,a}}$ can be defined by equation [4.7]

$$\sqrt{J_{2,a}} = \sqrt{\frac{1}{6}\begin{bmatrix} \left(S_{xx,a} - S_{yy,a}\right)^2 + \left(S_{yy,a} - S_{zz,a}\right)^2 \\ +\left(S_{zz,a} - S_{xx,a}\right)^2 + 6\left(S_{xy,a} - S_{yz,a} - S_{xz,a}\right)^2 \end{bmatrix}} \quad [4.7]$$

In this expression, the stress component $S_{\mathrm{ij},a}$ is defined by:

$$S_{\mathrm{ij},a} = \frac{1}{2}\left(\mathrm{max}_{0\leq t\leq T}\left[S_{\mathrm{ij}}(t)\right] - min_{0\leq t\leq T}\left[S_{\mathrm{ij}}(t)\right]\right);\ i,j = x,y,z \quad [4.8]$$

To simplify the calculation of $\sqrt{J_{2,a}}$, this transformation is commonly used [LAM 10]:

$$\sqrt{J_{2,a}} = \sqrt{D_{1,a}^2 + D_{2,a}^2 + D_{3,a}^2 + D_{4,a}^2 + D_{5,a}^2} = \|d_a\| \quad [4.9]$$

with

$$\begin{cases} D_1 = \frac{\sqrt{3}}{2}\left(\frac{2}{3}S_{xx} - \frac{1}{3}S_{yy} - \frac{1}{3}S_{zz}\right) \\ D_2 = \frac{1}{2}\left(S_{yy} - S_{zz}\right) \\ D_3 = S_{xy} \\ D_4 = S_{yz} \\ D_5 = S_{xz} \end{cases}$$

The methodology for calculating the average value of $\sqrt{J_{2,a}}$, in the frequency domain, is presented in the previous studies by [YAI 17]. As defined by equation [4.7], the estimation of the semi-axes R_i leads to the estimation of $\sqrt{J_{2,a}}$. The algorithm for determining $E\left[\sqrt{J_{2,a}}\right]$ is defined as follows:

1) Computation of the covariance matrix Σ_d of d defined by equation [4.10]. It can be determined from the stress covariance matrix Σ_s presented by equation [4.1]

$$\Sigma_d = E\left[\left(d(\mathrm{t}) - \bar{d}\right).\left(d(\mathrm{t}) - \bar{d}\right)^T\right] \qquad [4.10]$$

$$\Sigma_d = P\Sigma_s P^T \qquad [4.11]$$

2) The diagonal terms of Σ_d include the variance of each component. However, the diagonal terms show the variance between each component. In general, these terms are nonzero. Thereby, Σ_d is not diagonal. There exists an orthonormal basis of eigenvectors A_d such that:

$$\Sigma_{\tilde{d}} = A_d \Sigma_d {A_d}^T = diag(\sigma_1^2, \sigma_2^2, \sigma_3^2, \sigma_4^2, \sigma_5^2) \qquad [4.12]$$

with $\sigma_1 \geq \sigma_2 \geq \sigma_3 \geq \sigma_4 \geq \sigma_5$.

A vector d linearly associated with d by an orthogonal transformation can then be defined such that:

$$\tilde{d}(\mathrm{t}) = A_d d(t) \qquad [4.13]$$

with $\tilde{d}(\mathrm{t}) = [\widetilde{D_1}(t)\ \widetilde{D_2}(t)\ \widetilde{D_3}(t)\ \widetilde{D_4}(t)\ \widetilde{D_5}(t)]^T$.

3) Computation of the number of maxima $N_p(\widetilde{D}_\iota)$ for each parameter of the vector $d^{\sim}$. It can be determined from the spectral moment m_i of the random process $\widetilde{D}_\iota$:

$$N_p = \sqrt{\frac{m_4}{m_2}} \qquad [4.14]$$

$$m_i = \int_{-\infty}^{+\infty} |f|^i \Phi_{\widetilde{D}_\iota}(f) df \qquad [4.15]$$

4) $E[R_i]$ can then be estimated using equation [4.21], with $\sigma_{\widetilde{D}_\iota}$ introducing the standard deviation of the random process $\widetilde{D}_\iota$ [DAV 64]:

$$E[R_i] \approx \sigma_{\widetilde{D}_\iota}\left(\sqrt{2lnN_p\left(\widetilde{D}_\iota(t)\right)} + \frac{0,5772}{\sqrt{2lnN_p(\widetilde{D}_\iota(t))}}\right), i = 1 \ldots 5 \qquad [4.16]$$

5) For structures subject to random loading, $E\left[\sqrt{J_{2,a}}\right]$ can be determined by [PIT 01] and [BEN 08]:

$$E\left[\sqrt{J_{2,a}}\right] \approx \sqrt{E[R_1]^2 + E[R_2]^2 + E[R_3]^2 + E[R_4]^2 + E[R_5]^2} \qquad [4.17]$$

4.2.2. *Strategy for the fatigue damage analysis*

Figure 4.1 shows the implementation of the strategy of the fatigue damage analysis. The algorithm presented comprises two sub-problems: the first involves calculating the stress power spectral densities (PSDs). The model is first built by defining the geometry, material properties and loads. Next, a modal analysis is necessary to determine the normal modes of the model. Finally, by spectral analysis, the stresses PSDs are calculated at each structure element.

For the second subproblem, and in order to evaluate fatigue damage, the calculated PSDs are then input into the MATLAB© software to evaluate the stress covariance matrix first. This allows us to calculate the covariance matrix Σ_d. The covariance matrix $\Sigma_{\tilde{d}}$ and the PSD $\Phi_{\tilde{d}}$ are obtained by the orthogonal transformation of the matrix Σ_d.

The number of maxima N_p is calculated for each parameter of vector d. Next, the amplitude average of the shearing stress $E\left[\sqrt{J_{2,a}}\right]$ is determined. Finally, fatigue damage is determined at each structure point using the Sines criterion [YAI 17].

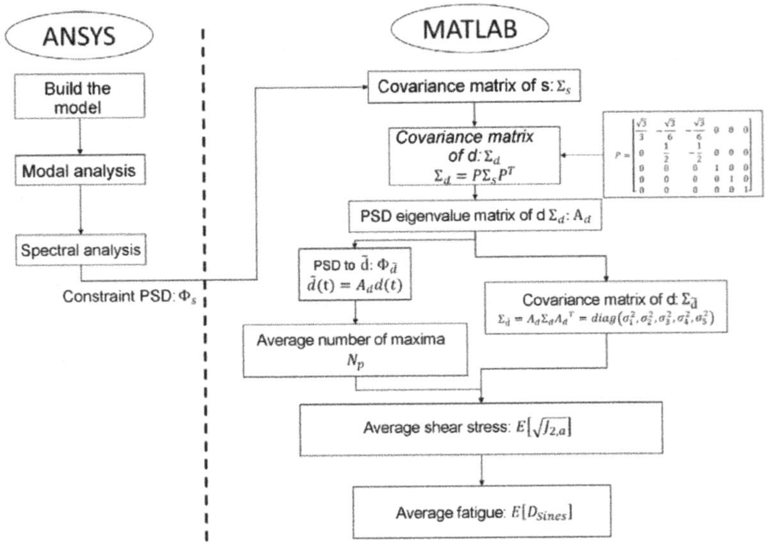

Figure 4.1. *Implementation of the strategy for the fatigue damage analysis. For a color version of this figure, see www.iste.co.uk/elhami/multi-physics.zip*

4.3. Reliability optimization of structures subjected to random vibrations

4.3.1. *Deterministic optimization*

The use of the deterministic optimization method (DDO) in structural design is becoming increasingly common in computer technology and the development of finite element analysis software. It consists of assigning safety factors to limit-state functions in order to take uncertainties in design variables into account.

Mathematically, the optimization problem can be defined by equation [4.19]. It then involves minimizing an objective function f (x) subject to a geometric, physical or functional constraint:

$$min_x f(x)$$

$$s.t\begin{cases} g_k(x) \leq 0\,, k = 1 \ldots K \\ h_j(x) = 0\,, j = 1 \ldots J \end{cases} \quad [4.18]$$

where x is the vector of the optimization variables, $f(x)$ is the objective function (also called the cost function), $g_k(x)$ and $h_j(x)$ are respectively the inequality constraints and the equality constraints. This approach may provide an optimal design, but the level of reliability cannot be controlled [KHA 09]. Reliability optimization is necessary for optimal and reliable designs.

4.3.2. *Reliability-based design optimization (RBDO)*

An RBDO problem can be thought of as an optimization problem under both deterministic and probabilistic constraints, which has a general form as expressed in the following equation:

$$min_x\, f(x)$$

$$s.t \begin{cases} P_r[G_i(x,y) \geq 0] \leq P_i^T \,, i = 1 \ldots m \\ h_j(x) \leq 0 \,, j = m+1 \ldots M \end{cases} \quad [4.19]$$

where x is the vector of the deterministic variables of the design, y is the vector of the random variables, $P_r[.]$ is the probability operator, G_i is the performance function, P_i^T is the target level of reliability and h_j is the deterministic constraint.

4.3.2.1. *Classical approach*

Traditionally, the reliability optimization procedure is performed within two spaces: the physical space and the normed space [ELH 13]. Therefore, the conventional approach of RBDO is achieved by integrating the following two problems.

Optimization problem

This problem consists of minimizing an objective function $f(x)$ under deterministic constraints $g_k(x)$ and reliability requirements. It can be defined by:

$$min_x\, f(x)$$

$$t.q \begin{cases} g_k(x) \leq 0 \,, k = 1 \ldots K \\ \beta(x,u) \geq \beta_t \end{cases} \quad [4.20]$$

where $\beta(x,u)$ is the structural reliability index and β_t is the target reliability index.

Reliability problem

$\beta(x,u)$ is the minimum distance between the limit-state function $H(x,u)$ and the origin in the normed space (see Figure 4.2). This index is calculated by the following minimization resolution:

$$\beta = min\ d(u)$$

$$t.q\ H(x,u) \leq 0 \qquad [4.21]$$

where $d(u)$ is the distance in the normed random space, given by $d(u) = \sqrt{\sum u_i^2}$.

Figure 4.3 depicts the algorithm of the conventional approach. It can be noted that this approach requires a large number of numerical computations.

Another more efficient approach called the hybrid method is presented; it is based on the simultaneous reliability and optimization solution and has been developed in [ELH 13].

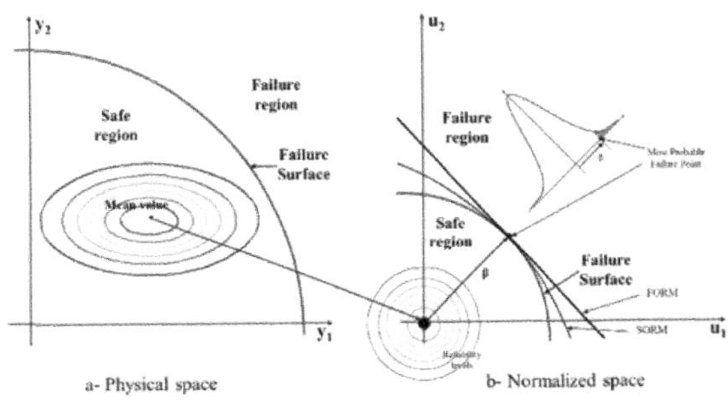

Figure 4.2. *Physical and normed spaces. For a color version of this figure, see www.iste.co.uk/elhami/multi-physics.zip*

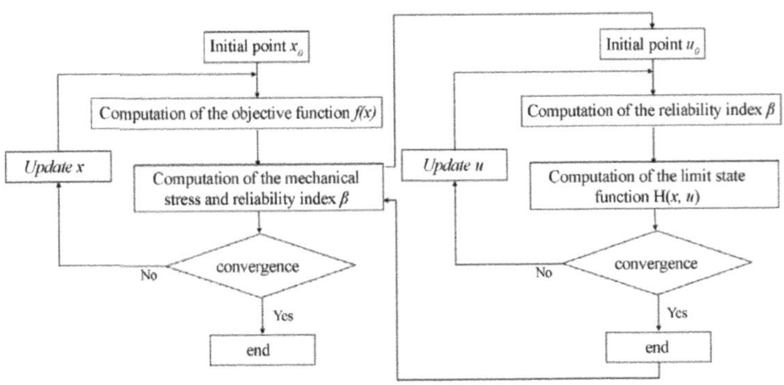

Figure 4.3. *Algorithm of the conventional approach to reliability optimization*

4.3.2.2. *Hybrid method*

This method consists of minimizing a new form of the objective function, named F(x, y) under limit-state constraints and deterministic constraints, as well as reliability constraints: this problem is written in the following form:

$$min_{x,y}\, F(x,y) = f(x) \times d_\beta(x,y)$$

$$s.t \begin{cases} G(x,y) \leq 0 \\ g_k(x) \leq 0\,, k = 1 \ldots K \\ d_\beta(x,y) \geq \beta_t \end{cases} \qquad [4.22]$$

Here, $d_\beta(x,y)$ introduces the distance, in the hybrid space, between the optimal point and the design point. The minimization of function F(x,y) is performed in the space (Hybrid Design Space (HDS)) of the deterministic variables x and the random variables y.

An example of HDS is given in Figure 4.4. For a normal distribution of random variables, d_β are represented by ellipses, the levels of the objective function are given by blue curves and the levels of the limit-state functions are represented by the black lines. Two important points can be seen: the optimal point P_x^* and the point of failure P_y^* (i.e. the point found on the curve G(x, y) = 0 and $d_\beta = \beta_t$). In the literature, the hybrid method effectively reduced the computational time compared to the conventional approach.

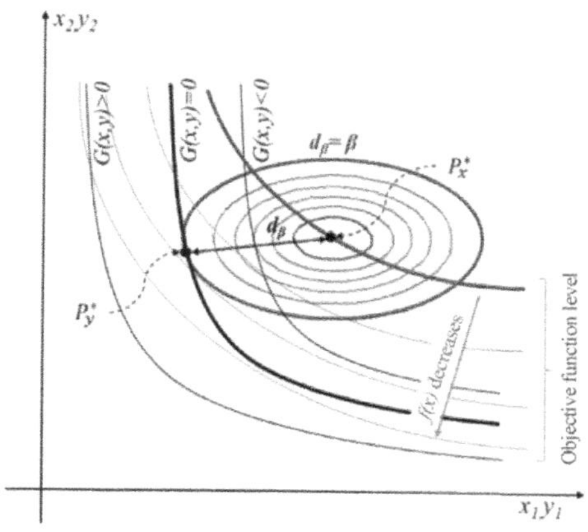

Figure 4.4. *Hybrid design space. For a color version of this figure, see www.iste.co.uk/elhami/multi-physics.zip*

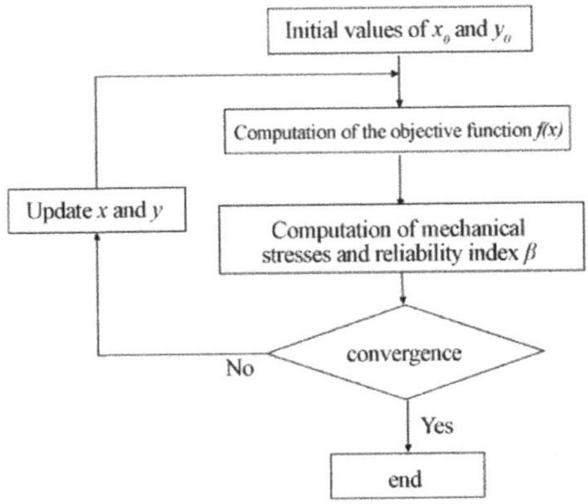

Figure 4.5. *Hybrid method algorithm*

Figure 4.5 shows the optimization algorithm of the hybrid method. The first step is to input the initial values of the random variables y0 and the deterministic variables x0. The second step involves evaluating the objective function F(x,y). Then, the optimization constraints as well as the reliability index are evaluated. The program stops if there is convergence and all optimization conditions are verified. Otherwise, it updates vectors x and y and repeats the operations from the second step.

4.3.2.3. *Optimum safety factor (OSF)*

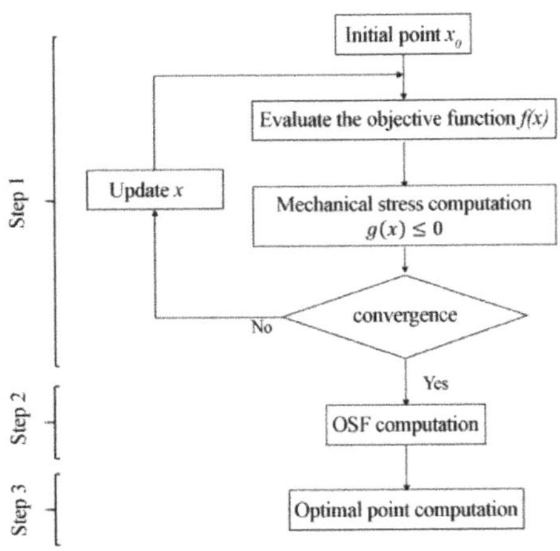

Figure 4.6. *OSF method algorithm*

In the deterministic optimization approach (DDO), the choice of the safety factor is based on engineering experience and the resulting solution cannot satisfy the appropriate level of reliability required. However, when designing a new structure, we usually need some experimental context to propose appropriate safety factors. With the OSF approach, the determination of safety factors should actually be independent of the engineer's experience.

Although the hybrid method leads to an improvement in the reliability level, the OSF approach leads to a minimization of computation time. It consists of using sensitivity analysis combined with reliability analysis in order to eliminate the disadvantages of the DDO method and the RBDO methods [ELH 13].

The OSF algorithm (see Figure 4.6) comprises three main steps:

Determination of the design point y_i^*

We minimize here the objective function under the limit-state function G(y) and the deterministic constraints. The resulting solution is designed by the most likely point of failure.

Computation of safety factors

In fact, the design point y_i^* is related to the optimal point x_i^* as defined in equation [4.22]. We consider here that the mean values m_i of the random variable are represented by the deterministic variable x_i^* [LOP 11]:

$$y_i^* = S_f^i\, x_i^*, i = 1, \dots, n \qquad [4.23]$$

Moreover, in general, for a normal distribution of the random variable y_i, we consider the transformation in the normed space (u-space) of the physical space (x-space) as defined in equation [4.23]:

$$u_i = \frac{y_i - x_i}{\sigma_i}, i = 1, \dots, n \qquad [4.24]$$

The safety factors can therefore be determined from u_i^* shown in equation [4.24] by combining equations [4.22] and [4.23]:

$$S_f^i = 1 + \gamma_i u_i^*, i = 1, \dots, n \qquad [4.25]$$

where γ_i is the coefficient relating the mean value and the standard deviation.

$$\gamma_i = \frac{\sigma_i}{m_i}.$$

According to [ELH 13], the optimal value of the normed vector u_i^* can be determined from the derivatives of the limit-state function at the level of the design point as follows:

$$u_i^* = \pm\beta_t \sqrt{\frac{\left|\frac{\partial G}{\partial y_i}\right|}{\sum_{j=1}^{n}\left|\frac{\partial G}{\partial y_j}\right|}}, i = 1, \dots, n \qquad [4.26]$$

with the sign $\pm$ depending on the sign of the derivative:

$$\frac{\partial G}{\partial y_i} > 0 \rightarrow u_i^* > 0 \text{ and } \frac{\partial G}{\partial y_i} < 0 \rightarrow u_i^* < 0, i = 1, \dots, n$$

Calculation of the optimal solution

In the final step, to determine the optimal design, we include the calculated safety factor values in the design variable value.

4.3.2.4. *Robust hybrid method (RHM)*

As previously said, the most significant disadvantage of the classic hybrid method is the high computation time. On top of that, it is not always stable and can converge toward an unfeasible solution. Alternatively, the optimal solution obtained P_x^* can be located in the failure region (G(x*)>0), while the distance between the optimal point and the design point P_y^* gives an acceptable level of reliability, as shown in Figure 4.7.

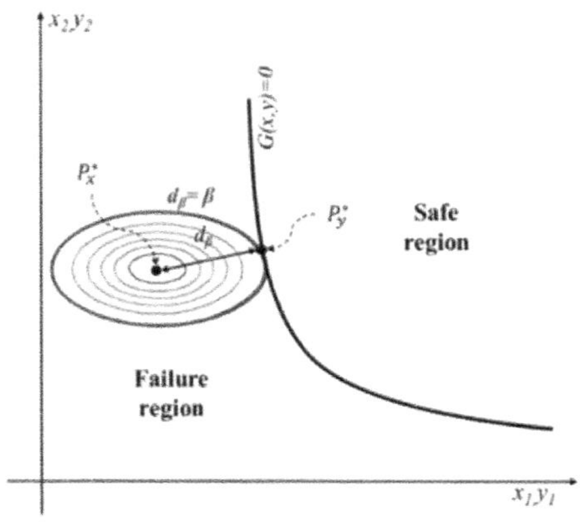

Figure 4.7. *MH issue [YAI 17]. For a color version of this figure, see www.iste.co.uk/elhami/multi-physics.zip*

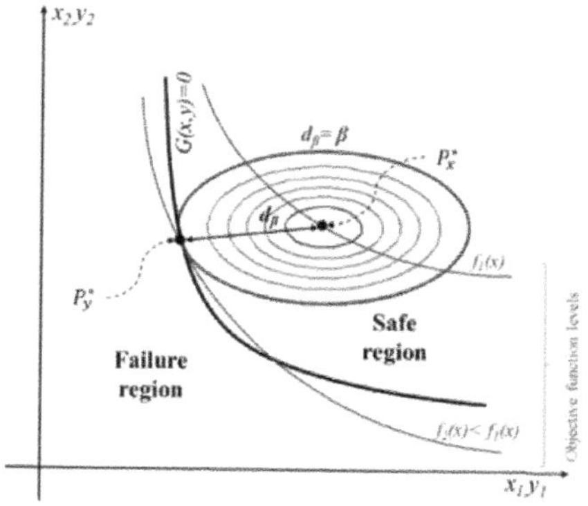

Figure 4.8. *RHM in the HDS [YAI 17]. For a color version of this figure, see www.iste.co.uk/elhami/multi-physics.zip*

A new method called the RHM is therefore developed to solve this problem. It consists of adding another constraint, to coerce the optimization algorithm to find the optimal solution in the safe region.

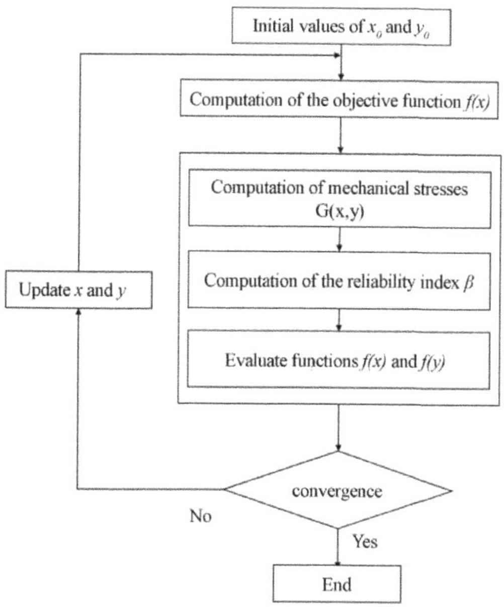

Figure 4.9. *RHM method algorithm*

Indeed, in the HDS, in order to ensure structural reliability under uncertainty, the value of the objective function at the optimal solution should be greater than this value at the design point. The RHM can then be defined by equation [4.26] [YAI 17].

$$min_{x,y} F(x,y) = f(x) \times d_\beta(x,y)$$

$$s.t \begin{cases} G(x,y) \leq 0 \\ g_k(x) \leq 0 \,, k = 1 \ldots K \\ \beta(x,u) \geq \beta_t \\ f(x) \geq f(y) \end{cases} \qquad [4.27]$$

Figure 4.8 depicts an example of the RHM method in the hybrid design space. f1(x) and f2(x) are the two levels of the objective function that intersect respectively with the optimal point P_x^* and the design point P_y^*. It can be observed that the optimal point P_x^* is located inside the safe region, f(x*) > f(y*) and that the required level of reliability is also met $(d_\beta = \beta_t)$.

The proposed method, whose algorithm is presented in Figure 4.9, not only allows the optimization algorithm to be forced into the hybrid design space, but it has also reduced the computation time compared to the classical hybrid method in an efficient way.

4.3.3. *Reliability optimization of structures subjected to random vibrations*

4.3.3.1. *Problem description*

Our goal now is to find the best design for a structure subjected to random vibrations. From the algorithm in Figure 4.1, it was possible to determine the average of the fatigue damage at each structure point.

Figure 4.10 shows the algorithm for optimizing structures subjected to random vibrations, in general. The first step involves evaluating the objective function from the initial values of the design variables. Then, the average of the fatigue damage is calculated at each structure point from the stress PSDs obtained by the finite element method.

The maximum of these values as well as the other optimization constraints are then evaluated. The program stops if there is convergence. Otherwise, the design variables are updated and the program repeats from the second step.

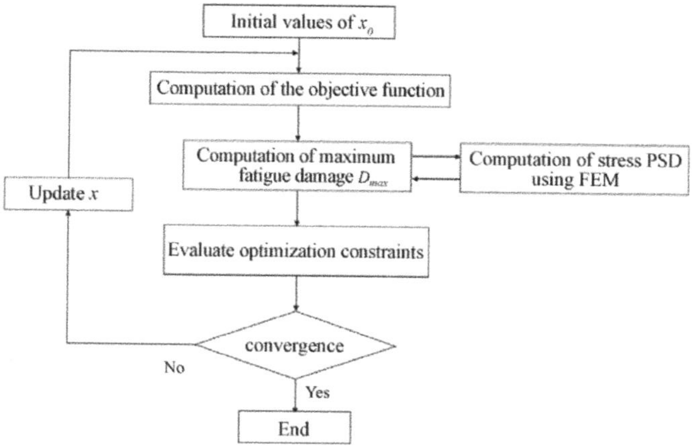

Figure 4.10. *Algorithm for optimizing structures subjected to random vibration*

4.3.3.2. *DDO and RBDO method procedures*

We denote by x the vector of deterministic design variables and y the vector of random design variables. The starting point (initial point) x_0, the design point y^* and the optimal point x^* can also be defined.

In the DDO approach, we consider a global safety factor S_f applied to damage D_{max} and based on engineering experience. This means that the optimal design of the structure must not exceed the maximal permissible values. Consequently, the DDO optimization problem can be defined as follows:

$$min_x\, f(x)$$

$$t.q\ D_{max}(x) \leq D_w \qquad [4.28]$$

The DDO procedure does not take into account the reliability constraint. Therefore, RBDO methods should be used.

Using the hybrid method problem, the objective function becomes a multiplication of the function $f(x)$ by the distance in the hybrid design space between the design point and the optimal point, as follows:

$$min_{x,y}\, F(x,y) = f(x) \times d_\beta(x,y)$$

$$t.q \begin{cases} D_{max}(y) \leq 1 \\ d_\beta(x,y) \geq \beta_t \end{cases} \qquad [4.29]$$

To solve the problems of the classic hybrid method, the RHM method is then proposed. It consists of adding a third constraint to the optimization problem. The optimization problem of the RHM is then:

$$min_{x,y}\, F(x,y) = f(x) \times d_\beta(x,y)$$

$$t.q \begin{cases} D_{max}(y) \leq 1 \\ d_\beta(x,y) \geq \beta_t \\ f(x) \geq f(y) \end{cases} \qquad [4.30]$$

To get the optimal solution by the OSF method, we need to determine the design point as the first step. This design point is obtained by solving the optimization problem defined by the following formulation:

$$min_y\, f(y)$$

$$t.q\ D_{max}(y) \leq 1 \qquad [4.31]$$

The safety factors are next calculated from the derivatives of the limit-state function D_{max}. The optimal solution is then calculated by dividing the values of the design points by the calculated safety factors.

4.4. Applications

4.4.1. *Problem description*

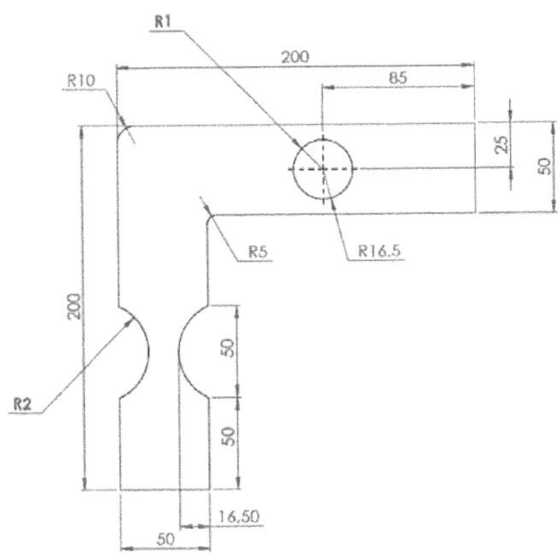

Figure 4.11. *Plate geometry*

We now consider the example of an L-shaped steel specimen whose geometry is defined in Figure 4.11. The specimen has a rounded notch and a hole, and is embedded at both ends. The plate thickness is 0.8 mm. Each support is subjected to a random acceleration whose PSD is a limited-band white noise that is exerted in the perpendicular direction to the plane of the specimen ($\Phi_a = 4\,(m/s^2)^2/Hz$; $f_c = 400\ Hz$). The accelerations of the two supports are completely independent.

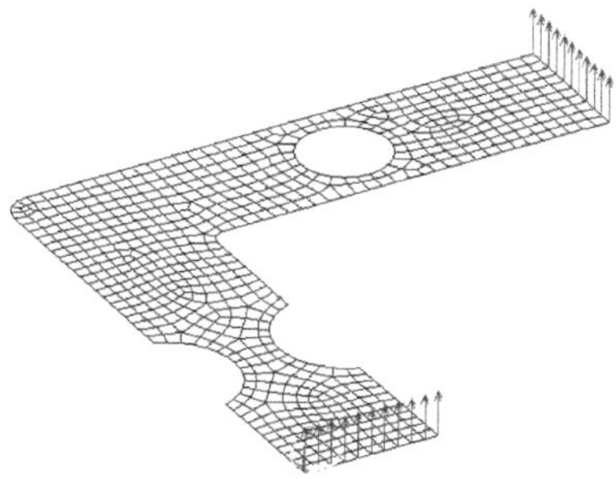

Figure 4.12. *Plate mesh. For a color version of this figure, see www.iste.co.uk/elhami/multi-physics.zip*

Figure 4.12 shows the finite element model and the boundary conditions. The arrows show the direction of random vibrations at each end of the structure. The plate has been discretized into 644 shell elements. Each element has four nodes with six degrees of freedom.

Figure 4.13 shows the first five modes that are included in the excitation band ($f_1 = 26.13\ Hz; f_2 = 101.88\ Hz; f_3 = 120.68\ Hz; f_4 = 274.42\ Hz; f_5 = 385.11\ Hz$).

The stress PSD matrix can be calculated in each element using a random vibration calculation code. The stress PSDs, $\Phi_{s_{xx}s_{xx}}$, $\Phi_{s_{yy}s_{yy}}$ and $\Phi_{s_{xy}s_{xy}}$ are shown in Figure 4.14, in a selected element (indicated in Figure 4.12).

Figure 4.15 shows the damage distribution calculated by the Sines criterion. This result is almost similar to the one based on the Crossland and Matake criteria [PIT 01].

The small difference between the results is caused by the mesh precision. The various procedures were implemented in MATLAB used as a post-processor of the ANSYS finite-element software.

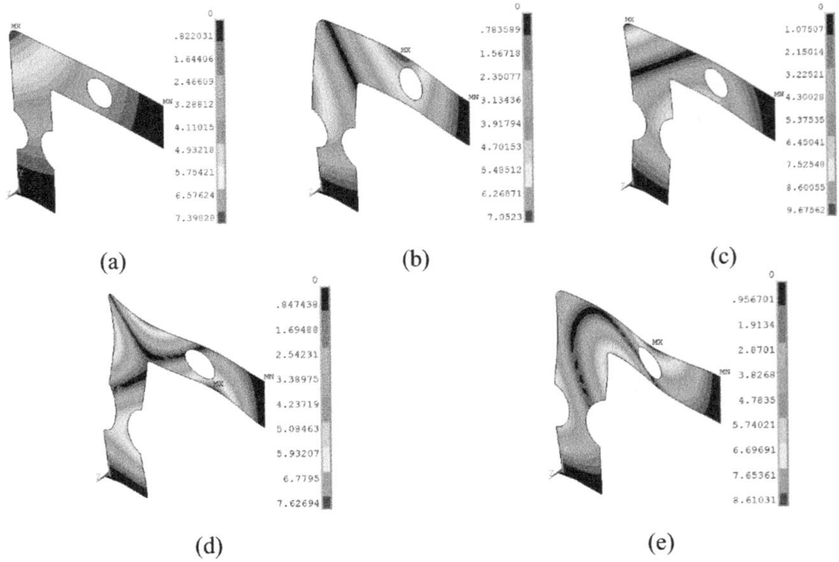

Figure 4.13. *Modal deformations of the plate: (a) Mode 1, (b) Mode 2, (c) Mode 3, (d) Mode 4 and (e) Mode 5. For a color version of this figure, see www.iste.co.uk/elhami/multi-physics.zip*

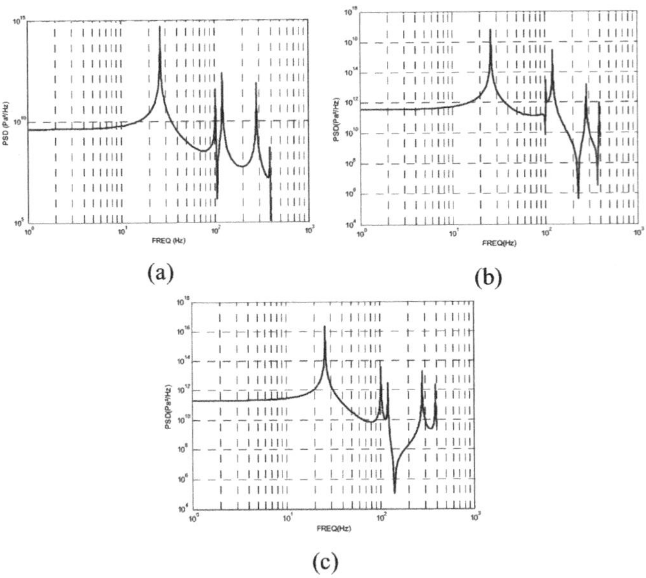

Figure 4.14. *Stress PSD: (a) $\Phi_{s_{xx}s_{xx}}$, (b) $\Phi_{s_{yy}s_{yy}}$ and (c) $\Phi_{s_{xy}s_{xy}}$. For a color version of this figure, see www.iste.co.uk/elhami/multi-physics.zip*

The objective now is to minimize the volume V of the part under the fatigue constraint D_{max}. In addition to the fatigue limit state, the distance d_β should be greater than or equal to the targeted reliability index β_t. In our case, β_t is equal to three, which corresponds to a failure probability virtually equal to 10^{-3}. The computational variables are the radius of the hole R_1, the radius of the notch R_2 and the plate thickness e. These design variables are normally distributed, and the variance coefficient γ of each variable is equal to 0.1; m_{R_1}, m_{R_2} and m_e are respectively the mean values of R_1, R_2 and e.

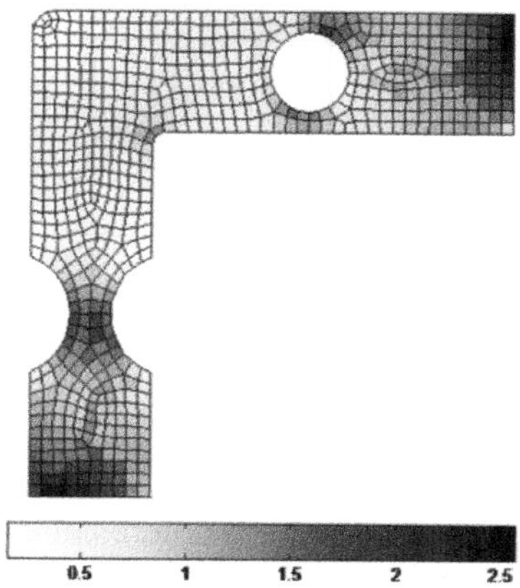

Figure 4.15. *Mapping of the Sines criterion fatigue damage. For a color version of this figure, see www.iste.co.uk/elhami/multi-physics.zip*

4.4.2. *Results and discussion*

In this section, we will present and compare the results of the different methods previously studied.

		Initial values x_0	Design point y^*	Optimal solution x^*
Design variable	R_1	15	17	16.53
	R_2	22	24.12	22.37
	e	2	0.96	1.06
Objective function	V	31 868.82	14 911.01	16 757.96
Optimization constraint	D_{max}	0.137	1	0.799
	β			1.27

Table 4.1. *Results of the DDO method*

Table 4.1 presents the results of the DDO method. The optimal solution satisfies the fatigue damage constraint but the level of reliability is lower than the target level of reliability β_t.

		Initial values y_0	Initial values x_0	Design point y^*	Optimal solution x^*
Design variable	R_1	16.5	15	16.58	17
	R_2	25	22	25.98	26
	e	0.8	2	1.17	0.9
Objective function	V	12357.91	31868.83	17957.41	13782.28
Optimization constraint	D_{max}	2.564	0.13	1.0001	**2.34**
	β		6.23		3

Table 4.2. *Results of the hybrid method*

The results of the hybrid method are presented in Table 4.2. It can be clearly seen that the resulting optimal point is in the failure domain ($D_{max} > 1$), whereas the distance between the optimal point and the design point β gives an acceptable level of reliability.

The results of the RHM method are presented in Table 4.3. We should note that the optimal solution is in the safe region and the level of reliability is verified. This change also contributed to minimize computation time. In fact, the computation time of the new formulation is about eight hours, while for the classic hybrid method, the computation time is about 27 hours without feasible result. The effectiveness of this new formulation is then demonstrated.

Table 4.4 shows the results of the OSF method. To find these results, a sensitivity analysis is required to show how various parameters affect the maximum fatigue damage. Table 4.5 defines the (derived) sensitivity of the maximum damage with respect to the different design variables.

With the OSF method, the computation time is about seven hours. This method allowed minimizing the computation time compared to the DDO method and the hybrid method. However, it is negligible compared to the RHM.

In addition, the value of the objective function of the optimal point obtained by the RHM is smaller than that found by the OSF method ($V_{OSF}^* > V_{RHM}^*$).

		Initial values y_0	Initial values x_0	Design point y^*	Optimal solution x^*
Design variable	R_1	16.5	16	17	14.16
	R_2	25	21	26	21.36
	e	0.8	1.5	1.16	1.11
Objective function	V	12357.91	23905.73	17829.22	17829.22
Optimization constraint	D_{max}	2.56	0.27	1	**0.77**
	β		5.05		3

Table 4.3. *Result of the RHM method*

To verify the effectiveness of the RHM method by changing the initial point, the optimization problem was solved with different initial points. It can be seen that, in most cases, RHM optimization converges toward a feasible solution. Table 4.6 shows the different initial points where the program converged. It should be noted that these points are chosen arbitrarily.

		Initial values x_0	Design point y^*	Normalized variables u^*	OSF S_f	Optimal solution x^*
Design variable	R_1	15	17	0.2271	1.023	16.62
	R_2	22	24.12	0.2009	1.02	23.64
	e	2	0.962	–2.9846	0.7	1.37
Objective function	V	31868.82	14911.01			21384.57
Optimization constraint	D_{max}	0.1372	1.0001			0.36
	β					3

Table 4.4. *Results of the OSF method*

$\frac{\partial D_{max}}{\partial R_1}$	$\frac{\partial D_{max}}{\partial R_2}$	$\frac{\partial D_{max}}{\partial e}$
0.017	0.013	–2.993

Table 4.5. *Maximum damage sensitivity analysis*

Figure 4.16 presents the initial objective functions V_{x_0} and V_{y_0} (black curves) and the optimal objective functions V_x^* and V_y^* (red curves) computed for each initial point. A brief analysis of these results shows that the RHM optimization problem converges nearly to the same optimal point (red curve). It can therefore be concluded that the RHM method is an efficient and robust method and can lead to the same optimal solution considering different initial points.

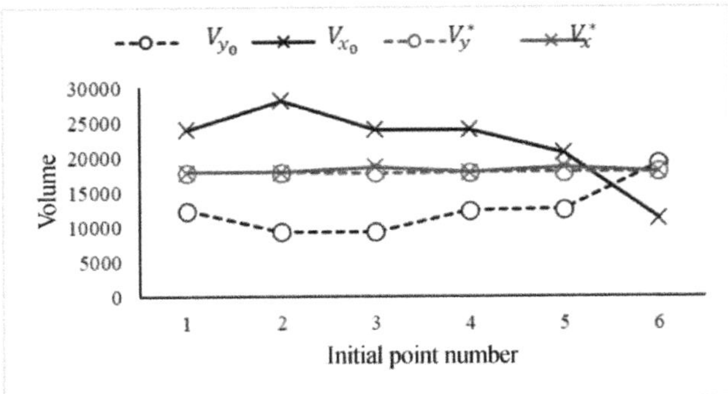

Figure 4.16. *Values of the objective function at the initial and optimal points for different initial points. For a color version of this figure, see www.iste.co.uk/elhami/multi-physics.zip*

		Point 1	Point 2	Point 3	Point 4	Point 5	Point 6
x_0	R_{10}^x	16	14	16	16	17	16
	R_{20}^x	21	26	21	21	21	20
	e_0^x	1.5	1.8	1.5	1.5	1.3	0.7
	V_{x_0}	23905.73	28010.47	23905.73	23905.73	20583.53	11221.6
	$D_{max}^{x_0}$	0.29	0.32	0.29	0.29	0.41	2.55
y_0	R_{10}^y	16.5	17	17	16.5	17	15
	R_{20}^y	25	24	24	25	23	22
	e_0^y	0.8	0.6	0.6	0.8	0.8	1.2
	V_{y_0}	12357.91	9308.38	9308.38	12357.91	12501.47	19121.29
	$D_{max}^{y_0}$	2.56	3.74	3.74	2.56	1.75	0.53

Table 4.6. *Different initial points for the RHM method*

Consequently, the RHM method provides the best results compared to other methods, for solving an RBDO problem involving structures subjected to multiaxial random vibrations. It can solve reliability problems compared with the DDO model and avoids the drawback of finding the optimal point inside the failure zone.

4.5. Conclusion

In this chapter, a method has been presented that results in obtaining an optimal design for structures subjected to random vibrations, thus ensuring a level of reliability, taking fatigue damage into account. We have introduced the three methods: OSF, DDO and RBDO.

The RHM was then introduced. An application with a structure subjected to random vibrations showed that the optimal solution found by this method is more reliable than the classical hybrid method and the DDO method. Using the RHM method, the objective function obtained is smaller than that obtained by the OSF method. In addition, the computation time of the RHM is almost the same as that of the OSF method. The effectiveness of the new method is then demonstrated.

5

Optimization and Shaping

5.1. Introduction

The hydroforming process involves several complex phenomena and often exhibits several types of nonlinearities (material, geometric, boundary conditions, etc.). Improving such an operation requires a lot of testing in order to accurately determine the loading paths and obtain defect-free parts. Digital tools have enabled manufacturers to simulate and optimize their means of production before launching the manufacturing, for the purposes of optimizing the number of defective parts. This is justified by the myriads of parameters to be controlled and also by the high cost of a real test. Several optimization methods have been proposed over the last ten years in order to properly perform a shaping operation. We thus mention the thesis by [BEN 11].

Most of these methods combine FEM (finite element method) and optimization techniques. Manufacturing processes can be virtually simulated. This step makes it possible to answer certain questions about the feasibility of the part and also about the ability of the loading path to correctly shape the part. This coupling leads to a significant improvement. However, it does not ensure the stability of the process or a certain type of plastic instabilities during manufacturing since there are several sources of uncertainty related to the material, loads, etc.

5.2. Different approaches to process optimization

We introduce the state-of-the-art of the main deterministic optimization techniques for optimizing simple and complex tubular shapes and also in the case of blank hydroforming. In [RAD 16], these techniques are classified into four main families:

– Conventional approach: in this section, the loading paths are determined on the basis of specific equations governing the flow of matter. Analytical methods are only valid for simple tubular forms. The limitations of these approaches reside in the fact that the equations cannot take into account certain physical phenomena. Among the works that have been based on analytical models, we have [ASN 03], which has validated the effectiveness of this approach by way of simulations (finite elements) and by experimental measurements. However, this approach turns out to be complicated for complex geometries.

– "Self-feeding" approach: this is proposed by [STR 01] and consists of making numerical simulations without taking into account axial forces or friction until the desired shape is obtained. In a second step, the displacement of the node located at the edge of the tube is measured and according to the thinning rate in the expanded area, the loading path is adjusted to include the appropriate axial displacement. This avoids the localization of the visualized thinning and subsequently the phenomenon of shrinkage. This approach has several disadvantages: it consists of performing several simulations for an adequate adjustment of the different parameters and consequently incurs a long computational time and a high cost. Figure 5.1 shows this approach. The measured displacement does not, in reality, take into account the friction effects between the die and the tube. This value is increased proportionally, after a new simulation is run and that includes the contact effects between the matrix and the tube. This approach seems inappropriate for nonsymmetrical parts such as T-shaped or Y-shaped parts, since the displacement rate is no longer the same on both sides.

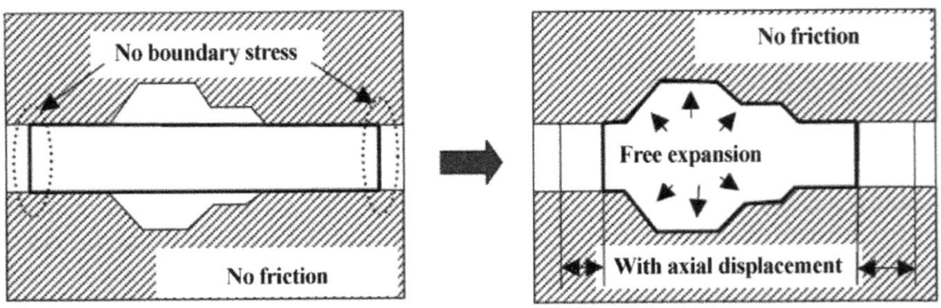

Figure 5.1. *Principle of the "self-feeding" approach*

– Adaptive approach: this consists of implementing a control algorithm for detecting the types of potential instabilities and able to react in order to accordingly read only the input parameters. This method has the advantage of converging towards the optimal solution in a single simulation. It results in reducing the computation time compared to the other techniques mentioned above. This technique has been adapted by several authors to optimize simple and complex tubular shapes. The adaptive approach is based on its ability to detect the onset of undesirable situations such as buckling, folding and necking followed by bursting. Depending on the type of plastic instability detected, the algorithm reacts by increasing or decreasing the internal pressure or axial displacement. [SHU 07] used an adaptive approach by combining it with a fuzzy logic-based technique to predict optimal loading paths. Criteria for each type of plastic instability are implemented for the detection of folding and necking. During each increment, the folding and necking values are evaluated and, depending on the values found, they adjust the new pressure and axial displacement parameters. In [JAN 08], an adaptive approach is also used: they rely on other types of criteria for the detection of folding and necking. For folding, they use an energy criterion based on bifurcation theory, while for necking they use the forming limit curve (FLC). Based on these two criteria, the parameters are adjusted during the simulation using a technique based on fuzzy logic. In [IMA 05], an adaptive optimization method based on a response surface technique is proposed. The adaptive approach can be summarized in Figure 5.2.

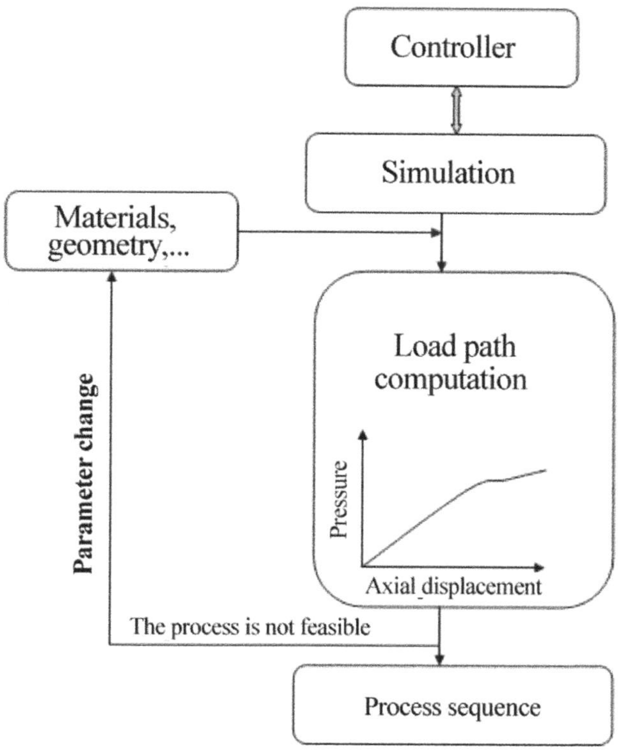

Figure 5.2. *Principle of the adaptive approach*

– Approach based on an optimization technique: this consists of defining an objective function to be optimized (minimized or maximized) with or without constraints. The criterion to be optimized may involve, for example, the distribution of the final thickness, the damage rate or the level of compliance of the part with respect to the die. An objective function [LOU 07] is defined which involves the variation of the thickness by imposing constraints on the final appearance of the part and also on the maximal permissible equivalent plastic stress. Several authors have adapted the same philosophy through the use of an objective function and constraints. In [FAN 03], the focus is on T-shaped tubes, which optimize the loading path by imposing two objective functions. The first concerns a minimization between the volume of the die and the final volume of the tube, while the second ensures a uniform distribution of the final thickness. In their problem, they impose a constraint that controls the equivalent stress in each element that must not exceed the ultimate stress of the material. The conjugate gradient method is used to optimize a T-shaped tube [ELH 13]. Other techniques have been proposed, such as that of [BEN 14], which couples the FEM with an intelligent optimization technique combining a local search technique (response surface or quadratic method) and a global search technique (genetic algorithm). In recent years, the response surface method has been widely used for the optimization of forming processes. It has proven to be a useful tool to mitigate issues with computational time while ensuring a good accuracy of the final result.

Deterministic optimization techniques have some drawbacks since they do not take into account process-related variabilities. These variabilities can be caused by: material parameters, the geometry of the part, loading parameters, lubrication conditions, as well as external factors such as temperature or humidity effects and tool wear. Failure to take these variabilities into account can lead to the appearance of certain types of plastic instabilities. The deterministic approach therefore does not guarantee the stability, nor the robustness of the process with regard to the failure modes that may appear, and it does not take into account the uncertainties of certain parameters. Their integration into the optimization process is essential to ensure an acceptable level of reliability. This makes it possible to significantly reduce the rate of defective parts. In order to take these variabilities into account, new approaches have been proposed that allow for the integration of these uncertainties, such as the reliability approach or the robust approach [BEN 13].

These approaches appear to be promising alternatives for both optimizing the process and also ensuring an acceptable level of reliability. However, the accuracy and reliability of the FEM is highly dependent on the precision of the input parameters. In reality, these parameters must be defined by their statistical moments and their probability distributions. Access to this information is expensive and requires a test campaign for the characterization of the material or with the variability of the machine. For this reason, our probability laws as well as coefficients of variation are defined in the form of hypotheses.

5.3. Characterization of forming processes by objective functions

The optimization of forming processes is generally done by defining objective functions and constrained functions. These functions are used to control the process and ensure good quality of the final product during the process. They also make it possible to avoid certain failure modes that may appear. Several studies that have appeared in the literature have been inspired by the FLC, whether at the level of stresses or main strains, to define their functions. Some define functions that involve the level of compliance with respect to the matrix in order for the part to take the shape of the matrix. Other types of functions that can be defined consists of minimizing the rate of damage in the room.

The optimization of forming processes by way of some objective functions with or without deterministic constraints is a widely used approach. Nonetheless, it is often criticized because it does not take into account the uncertainties on the parameters even though these hazards are present. Throughout this chapter, the hydroforming process is optimized with different types of objective functions by replacing deterministic constraints with probabilistic constraints. The coupling between mechanical and probabilistic aspects is also presented.

5.4. Deterministic and probabilistic optimization of a T-shaped tube

This example has been widely studied in the literature but namely for the topic of deterministic optimization. The problem is solved in a deterministic way then with a probabilistic approach in order to compare the two and emphasize the contribution of the probabilistic approach. The same methodology was presented for tube hydroforming [RAD 10; AYA 11]. The formulation of the optimization problem can be outlined in different ways depending on the objectives sought after.

A finite element model is defined to perform all of the necessary numerical simulations. This model is composed of a die that is used to give the tube its shape and a punch whose function is to feed material to the expanded area in order to prevent premature necking of the tube. Both of these bodies are modeled by rigid bodies. The tube is 121 mm long, with a diameter of 45.4 mm and 1.2 mm thick initially. The finite element model and the mesh are given in Figures 5.3 and 5.4. Given the symmetry of the problem and in order to reduce the computation time, a quarter of the model is modeled.

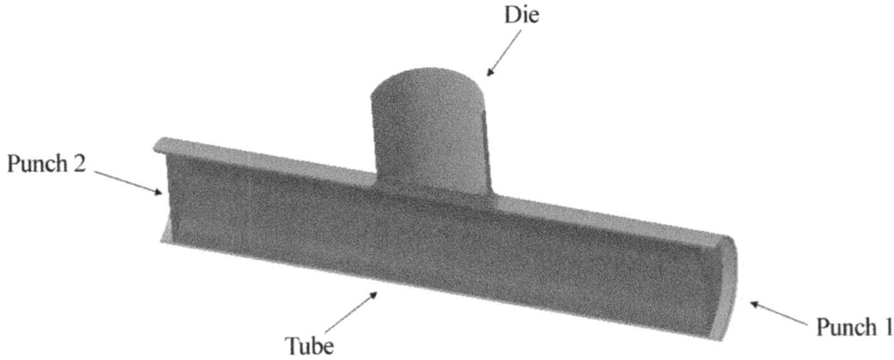

Figure 5.3. *Finite element model: exploded view. For a color version of this figure, see www.iste.co.uk/elhami/multi-physics.zip*

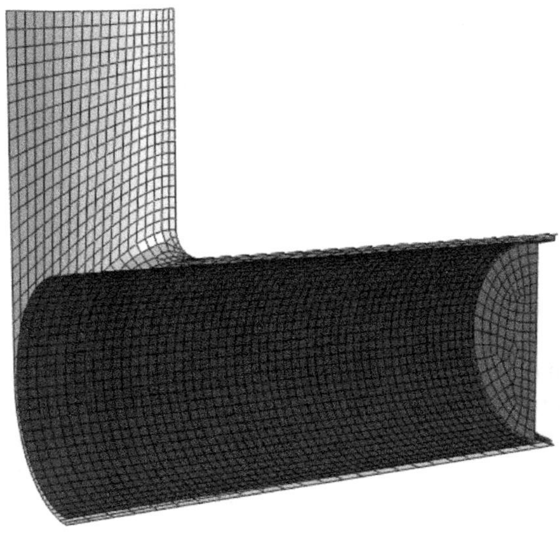

Figure 5.4. *Adapted model mesh. For a color version of this figure, see www.iste.co.uk/elhami/multi-physics.zip*

The material used for the shaping of this tube is DC06 steel. A Swift work hardening model is chosen, the mechanical properties of which are presented in Table 5.1.

k (MPa)	n	ε_0
527	0.24	0.0024

Table 5.1. *Work hardening model coefficients*

5.4.1. *Selection of objective function and definition of constraints*

The resolution of this type of problem can be formulated in different ways. For instance, the goal at the onset would be to ensure uniformity in the distribution of the final thickness, a minimization of the damage rate, etc. The main point here is to maximize the height of the dome. We denote by u_{max} the maximal height reached during the process. The problem constraints are defined to avoid probable failure modes during the process.

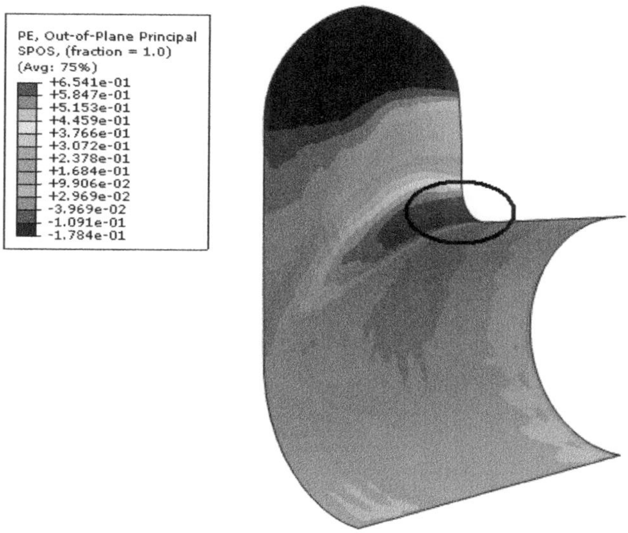

Figure 5.5. *Localization and limit value of out-of-plane strain. For a color version of this figure, see www.iste.co.uk/elhami/multi-physics.zip*

Tube hydroforming often involves two types of plastic instabilities that have to be controlled: folding and necking. Two criteria are defined to avoid that these two failure modes occur. The evaluation and control of these two plastic instabilities leads to obtaining a defect-free part. These two criteria are illustrated as follows:

– Folding criterion: this is based on the critical out-of-plane strain. Indeed, the maximal value of this strain is located at the level of the transition zone as shown in Figure 5.5. A high strain value is synonymous with folding. This critical value of deformation is equivalent to $\tilde{\varepsilon}_{hp} = 0.65$, determined using some deterministic simulations.

– Necking criterion: this is based on a simple criterion which consists of defining a critical thinning rate that should not be exceeded. In this case, a thinning rate of approximately 25% is tolerated.

5.4.1.1. *Selection of uncertain parameters*

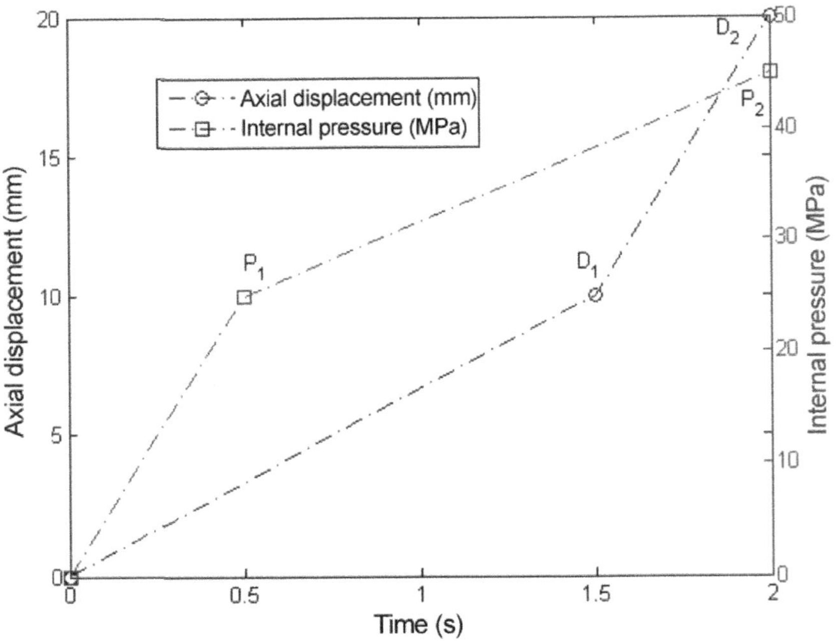

Figure 5.6. *Loading paths. For a color version of this figure, see www.iste.co.uk/elhami/multi-physics.zip*

The choice of loading parameters as uncertain parameters is justified by the fact that the process exhibits a certain sensitivity to these parameters. In fact, a slight variation of these parameters around their average values can actually lead to the process failing and the occurrence of plastic instabilities [HUA 16a]. The success of this type of process requires good synchronization between the internal pressure and the axial displacement applied. Taking these uncertainties into account in the optimization allows us to better determine the loading routes that make the process resistant to different variations. The material parameters as well as the coefficient of friction between the die and the tube are considered deterministic defined by their nominal values.

The loading paths giving the change in pressure and axial displacement as a function of time are shown in Figure 5.6. These loading parameters are given according to their average values. The probabilistic characteristics of these parameters are presented in Table 5.2.

V.a.r.	Average	Var. coef. (%)	Distribution
P_1	25	5	Normal
P_2	45	5	Normal
D_1	10	5	Normal
D_2	20	5	Normal

Table 5.2. *Probabilistic characteristics of loading parameters*

5.4.1.2. *Construction of the objective function and constraints*

The solution of the optimization problem is based on the construction of metamodels giving the objective function and the constraints according to the uncertain parameters. A centered composite design of experiment is chosen to perform all of the necessary simulations. Given the nonlinearities present in the problem, a quadratic model is chosen to represent all of these responses. Figures 5.7–5.9 represent the maximum displacement at the dome, maximal out-of-plane strain and the final thinning at the dome, respectively. A statistical analysis of metamodels shows that the latter provide a fairly acceptable level of accuracy for predicting responses. The coefficients of determination are presented in Table 5.3.

Response	u_{max}	ε_{hp}	Thinning
R^2	0.997	0.9393	0.9954
R^2_{adj}	0.9856	0.7175	0.978

Table 5.3. *Statistical indicators*

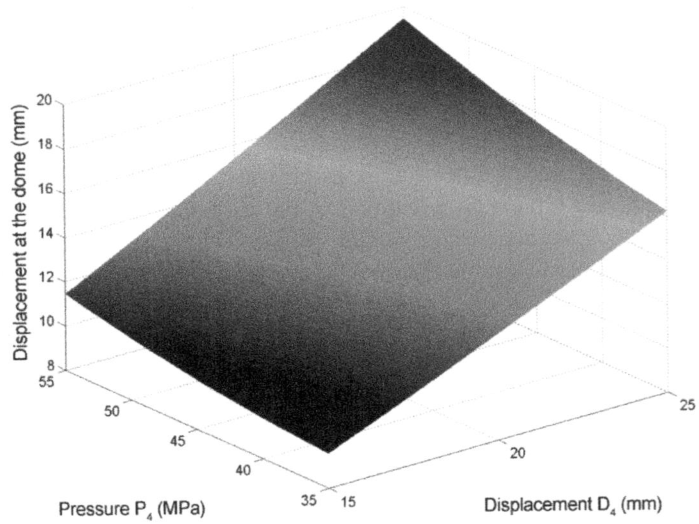

Figure 5.7. *Maximal displacement approximation. For a color version of this figure, see www.iste.co.uk/elhami/multi-physics.zip*

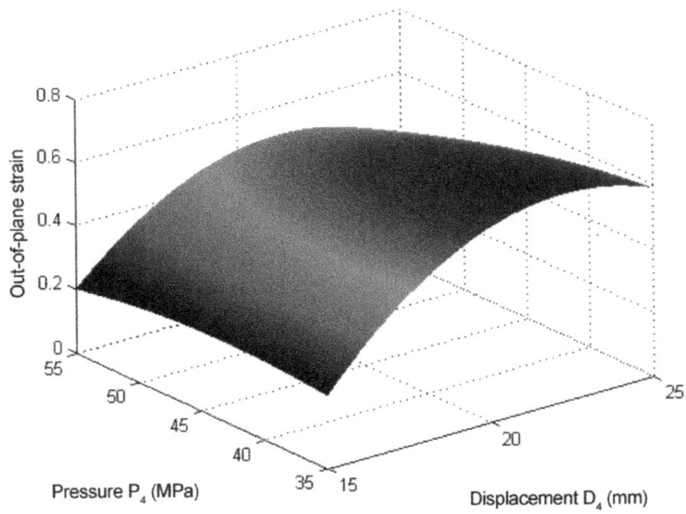

Figure 5.8. *Out-of-plane strain approximation. For a color version of this figure, see www.iste.co.uk/elhami/multi-physics.zip*

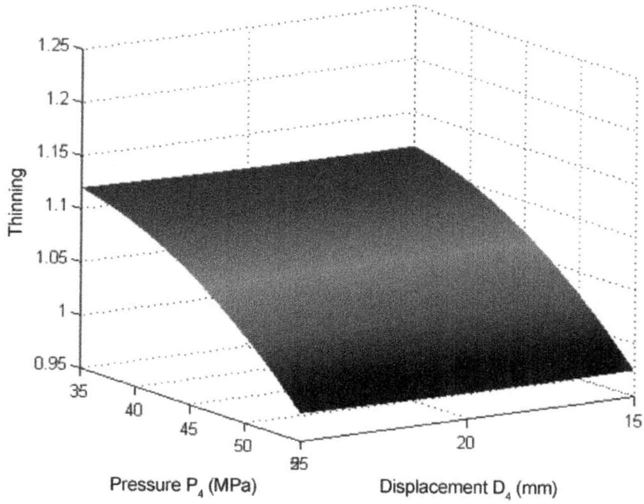

Figure 5.9. *Thinning approximation. For a color version of this figure, see www.iste.co.uk/elhami/multi-physics.zip*

5.4.2. *Deterministic formulation of the optimization problem*

The problem is first of all solved in a deterministic way. Two types of nonlinear optimization algorithms are used to determine the deterministic optimum: a gradient-based algorithm and a "pattern search (PS)" directional search algorithm based on a grid technique for the search for the optimum. Mathematically, the deterministic optimization problem can be formulated as follows:

$$\begin{cases} \max_{g_1(x_i)} u_{max}(x_i) \\ g_1(x_i) = \tilde{\varepsilon}_{hp} - 0.65 \leq 0 \\ g_2(x_i) = 0.95 - \tilde{h}_{min} \leq 0 \\ x_i = \{D_1, D_2, P_1, P_2\} \end{cases} \qquad [5.1]$$

The optimization problem involves maximizing the dome height while ensuring that neither necking nor folding will occur during shaping. For controlling the necking, a thinning of the order of 25% is tolerated.

Solving the problem with an SQP-type algorithm makes it possible to reach the deterministic optimum after 19 iterations. The optimum loading paths obtained in the deterministic case are compared with the exiting paths in Figure 5.10. The maximal dome height is also shown in the deterministic case, which reaches a maximum value of 19.457 mm (see Figure 5.11).

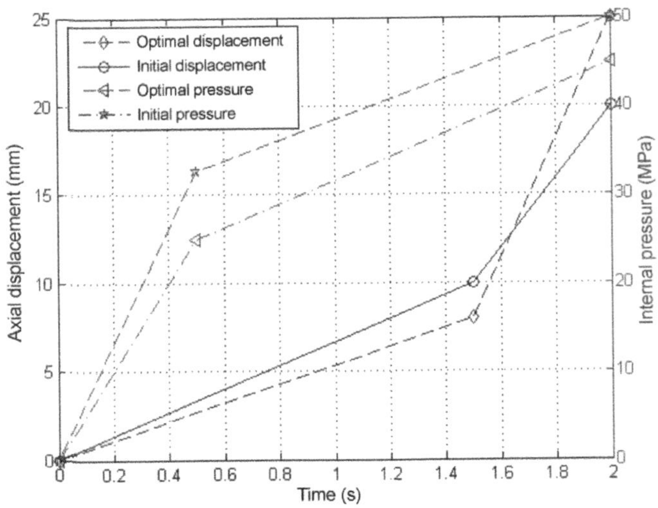

Figure 5.10. *Comparison of deterministic optimum paths with initial paths. For a color version of this figure, see www.iste.co.uk/elhami/multi-physics.zip*

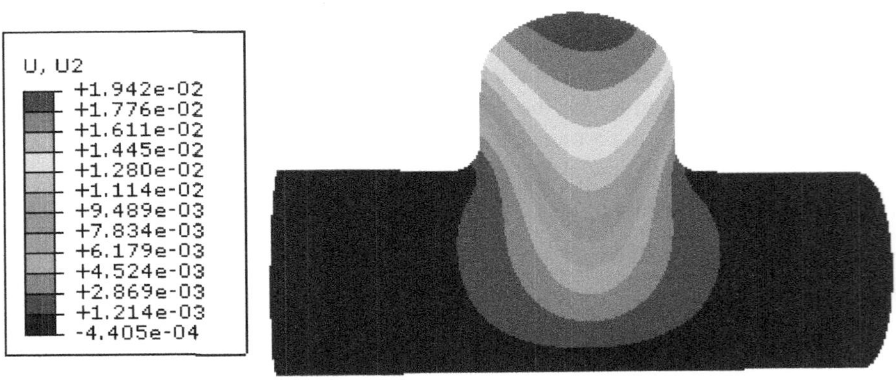

Figure 5.11. *Maximum height at the dome. For a color version of this figure, see www.iste.co.uk/elhami/multi-physics.zip*

With the PS algorithm, the algorithm converges after five iterations and 643 evaluations of the objective function. The convergence path of these two algorithms is given in Figure 5.12.

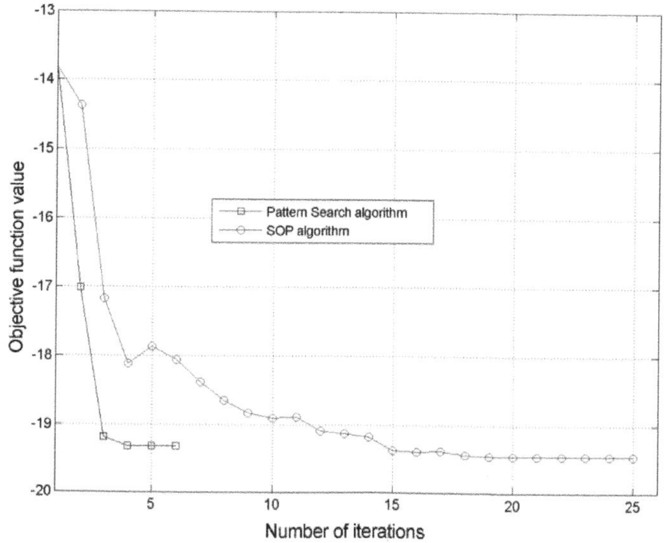

Figure 5.12. *Convergence path for the SQP algorithm and the PS algorithm. For a color version of this figure, see www.iste.co.uk/elhami/multi-physics.zip*

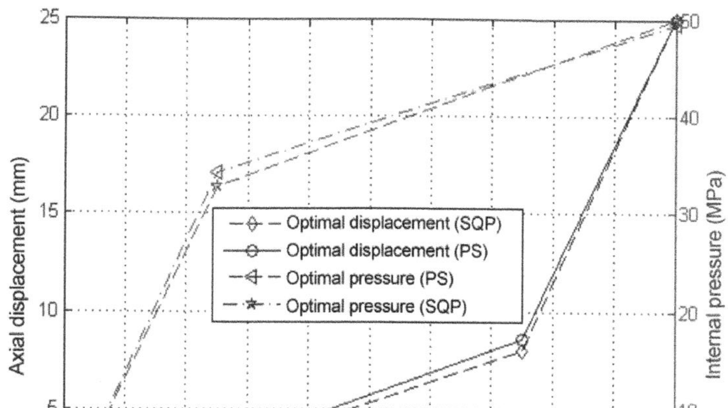

Figure 5.13. *Comparison of the optimum paths obtained with the two algorithms. For a color version of this figure, see www.iste.co.uk/elhami/multi-physics.zip*

The optimum loading routes obtained with this algorithm are compared with those obtained by the algorithm (SQP). The maximum dome height obtained with the PS algorithm is lower than that obtained with the SQP algorithm: it is 19.3227 mm.

Figure 5.13 shows the distribution of displacement for this second case. The use of these two types of algorithms shows that the SQP algorithm converges to a lower value, which proves its effectiveness compared to the PS algorithm. On the other hand, the latter converges faster towards the optimal solution.

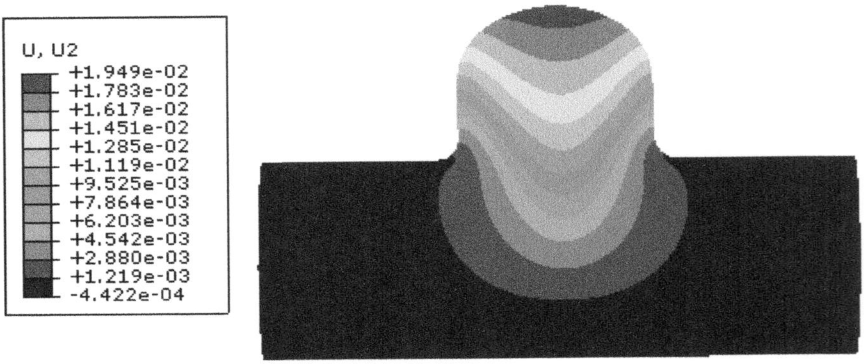

Figure 5.14. *Maximum height at the dome. For a color version of this figure, see www.iste.co.uk/elhami/multi-physics.zip*

5.4.3. *Probabilistic formulation of the optimization problem*

In the following, the problem is solved following a reliable method. The deterministic constraints formulated in equation [5.1] are now transformed into probabilistic constraints that involve the evaluation of the probability that one of these constraints will not be satisfied. It should also be noted that it is assumed that these two failure modes are independent, which means that the occurrence of one does not depend on the other. Within a reliability context, the constraint is defined by a limit state function or performance function. The two limit state functions are expressed according to the deterministic constraints, and they are given by the system of equations in [5.2]:

$$\begin{cases} G_1 & = -g_1 \\ G_2 & = -g_2 \end{cases} \qquad [5.2]$$

The reliability problem consists of optimizing the average values of the optimization variables given by the vector $X = (\mu_{D1},\mu_{D2},\mu_{P1},\mu_{P2})$. The reliability optimization problem is formulated as follows:

$$\begin{cases} \max_{[G_1(x)]}(\mu_i) \\ P[G_1(x) \leq 0] \leq P_{\text{admissible}} \\ P[G_2(x) \leq 0] \leq P_{\text{admissible}} \\ \mu_i = \{\mu_{D_1}, \mu_{D_2}, \mu_{P_1}, \mu_{P_2}\} \end{cases} \quad [5.3]$$

The optimum loading paths are sought for different levels of permissible reliability, P_f = 2.28%, P_f = 0.62% and P_f = 0.13%, which correspond to the following reliability indices: 2, 2.5, 3 and 4. In the case of reliability optimization, the SQP algorithm is used to solve the problem. The algorithm converges for the different levels of reliability. The convergence of the objective function according to the number of iterations for these different cases is shown in Figure 5.15.

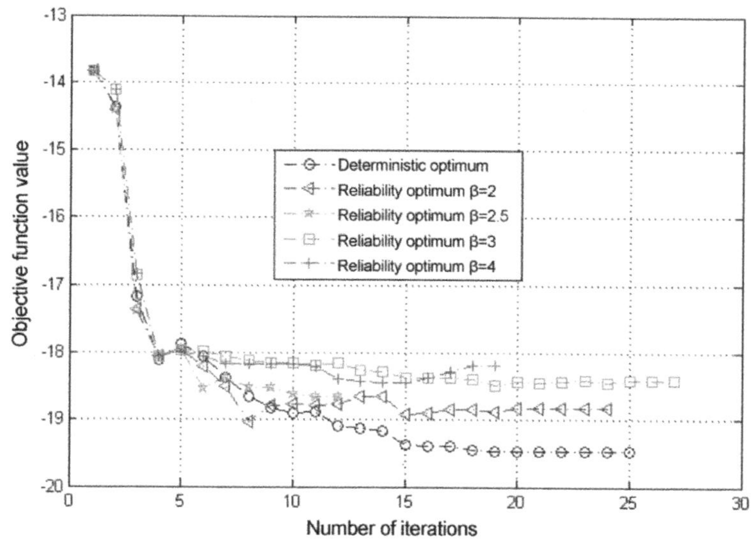

Figure 5.15. *Evolution of the objective function during iterations. For a color version of this figure, see www.iste.co.uk/elhami/multi-physics.zip*

The loading paths obtained for the different levels of reliability are compared with the outbound paths and the optimal deterministic paths given in Figures 5.16 and 5.17.

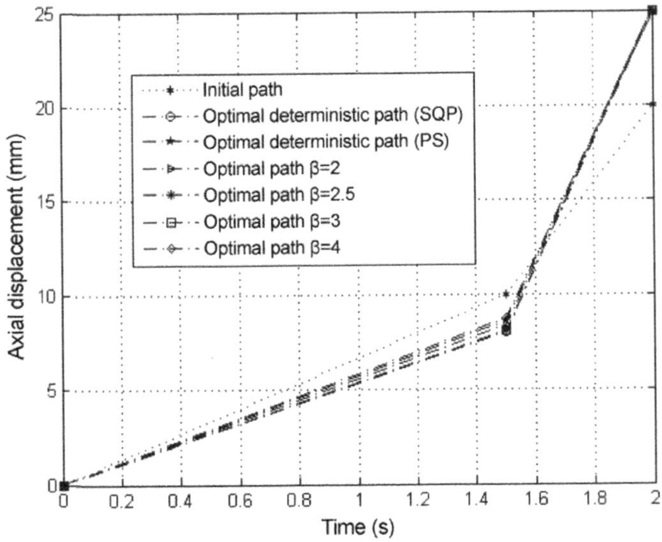

Figure 5.16. *Comparison of optimum paths: axial displacement (mm) – time (s). For a color version of this figure, see www.iste.co.uk/elhami/multi-physics.zip*

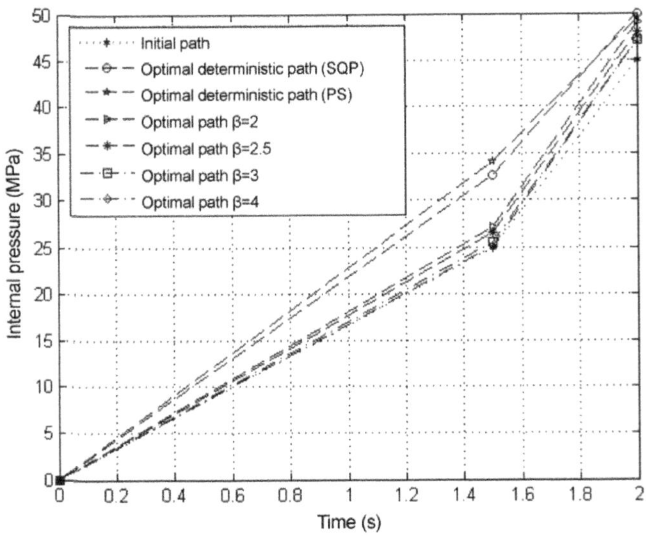

Figure 5.17. *Comparison of optimum paths: internal pressure (MPa) – time (s). For a color version of this figure, see www.iste.co.uk/elhami/multi-physics.zip*

The maximal displacement at the dome for the different levels of reliability is given in Figures 5.18–5.21. The higher the level of reliability, the lower the dome height compared to the deterministic case.

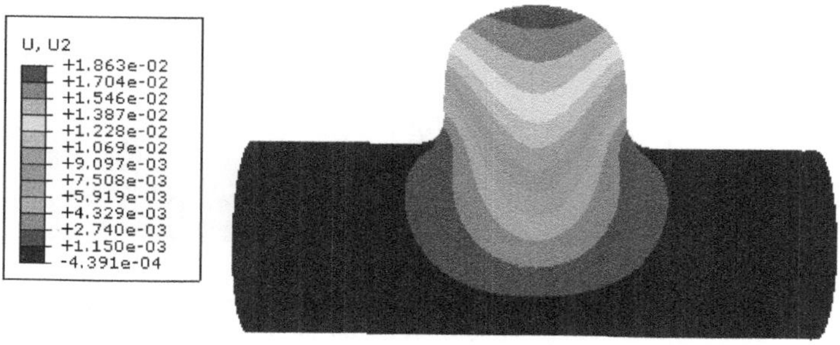

Figure 5.18. *Maximal displacement for $\beta = 2$. For a color version of this figure, see www.iste.co.uk/elhami/multi-physics.zip*

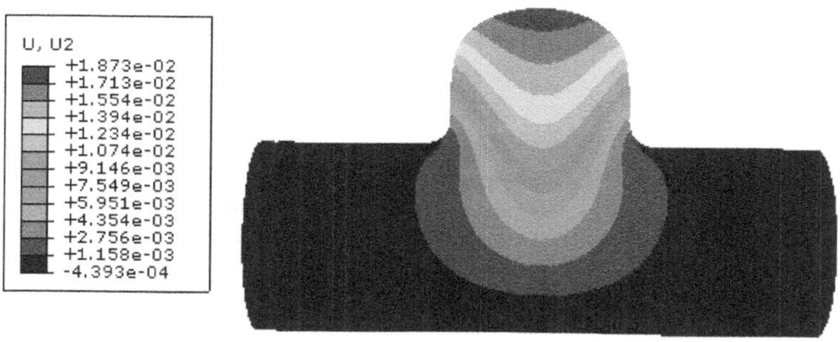

Figure 5.19. *Maximal displacement: $\beta = 2.5$. For a color version of this figure, see www.iste.co.uk/elhami/multi-physics.zip*

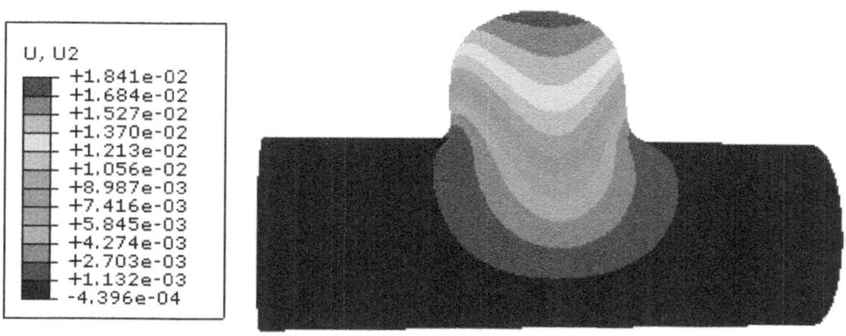

Figure 5.20. *Maximal displacement: $\beta = 3$. For a color version of this figure, see www.iste.co.uk/elhami/multi-physics.zip*

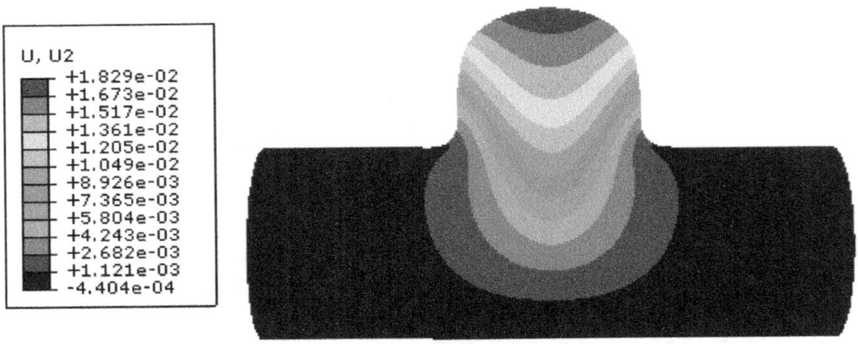

Figure 5.21. *Maximal displacement: $\beta = 4$. For a color version of this figure, see www.iste.co.uk/elhami/multi-physics.zip*

The thickness distribution for the deterministic optimum and the reliability optimum is shown in Figure 5.22.

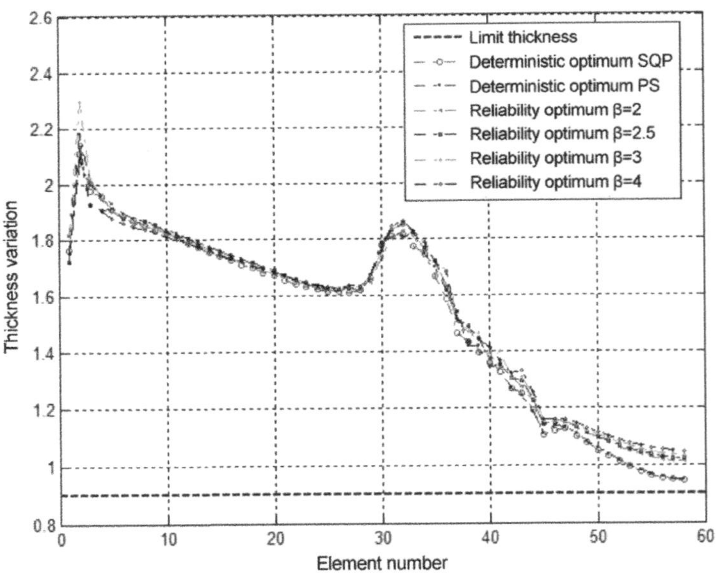

Figure 5.22. *Thickness variation according to the axial position. For a color version of this figure, see www.iste.co.uk/elhami/multi-physics.zip*

Figure 5.23 shows the height at the dome. In comparison with the thickness distribution given in Figure 5.22. It is noticeable that the greater the height of the dome, the greater the thinning.

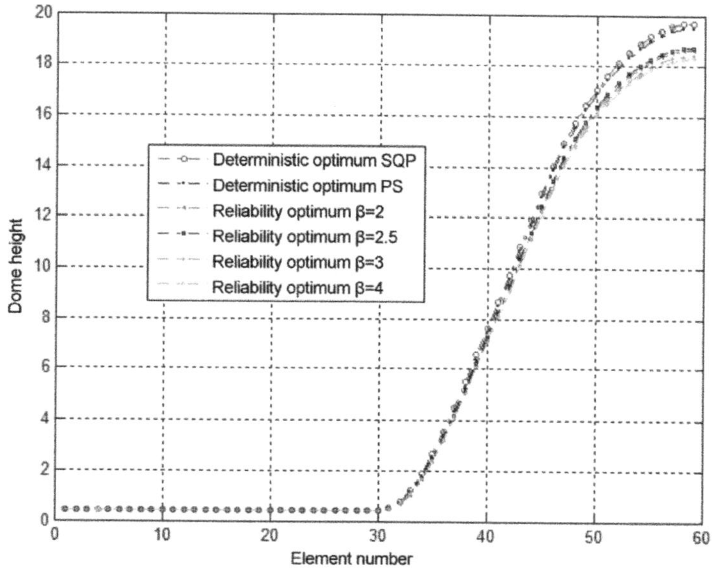

Figure 5.23. *Height at the dome for the various optimums. For a color version of this figure, see www.iste.co.uk/elhami/multi-physics.zip*

The localization of the optimums is given in the space of the parameters that have a great influence, namely (P_2, D_2) determined with a sensitivity analysis. The outline of the objective function and the constraints are plotted in Figure 5.24. The interest of reliability optimization resides in the search for the optimum that satisfies a targeted level of reliability.

In fact, as shown in Figure 5.24, the deterministic optimum always tends towards the limits of the constraints. By taking into account the uncertainties that affect certain parameters, these can incite the appearance of plastic instabilities. The interest of reliability optimization is better understood with this example when the target reliability index is increased. For example, it can be seen that the optimum that ensures a level of reliability equivalent to a reliability index $\beta = 4$ is much further from the deterministic optimum and the limits of the constraints. With this level of reliability, we move away from the risk area, which results in ensuring stability in the process [ELM 15].

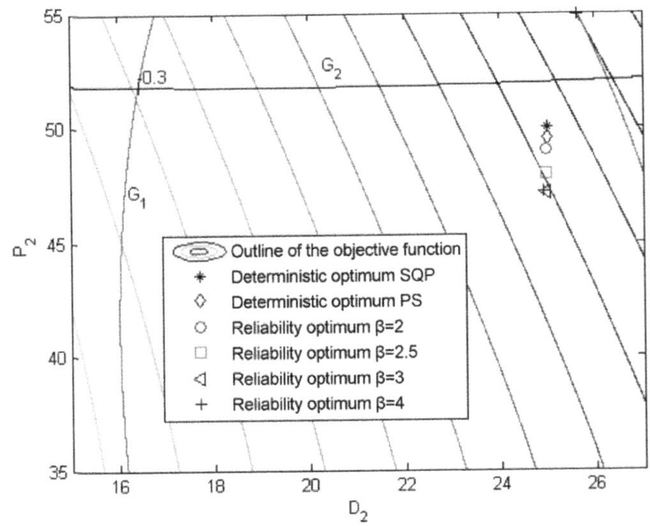

Figure 5.24. *Localization of optimums in the search space. For a color version of this figure, see www.iste.co.uk/elhami/multi-physics.zip*

In order to carry out further investigations on the influence of the starting point on the convergence towards the reliability optimum. The latter is searched for different starting points, we merely consider the case where the reliability index is set at $\beta = 3$. Table 5.4 summarizes the optimums found for six different starting points and also gives the number of iterations as well as the value of the objective function. As a matter of fact, forming problems are strongly nonlinear and converge towards different local optimums by changing the initial point. In this type of problem, several starting points are often tested and among all of these optimums, the one with the lowest objective function value is chosen.

Starting point	Reliability optimum	Number of iterations	Value of the f_{obj}
[10 20 25 45]	[8.09 24.95 25.45 47.13]	27	−18, 41
[11 18 23 47]	[8.85 24.88 24.04 48.28]	10	−18, 29
[9 22 27 43]	[8.18 24.97 27.95 46.20]	20	−18, 38
[11 22 24 49]	[8.34 24.97 22.55 48.53]	26	−18, 46
[12 20 21 46]	[8.34 24.97 26.33 47.93]	34	−18, 58
[9 19 21 47]	[9.46 24.97 22.39 49.85]	20	−18, 37

Table 5.4. *Influence of the starting point on the reliability optimum*

5.4.4. *Sensitivity of optimums to uncertainties*

In order to study the sensitivity of the deterministic and reliable optimum to uncertainties, these two optimums are perturbed to observe their ranges of variation. For the reliability optimum, the one that corresponds to a reliability index $\beta = 4$ is chosen. In Figure 5.25, it can be clearly seen that the reliability optimum does not exceed the limits imposed by the two stresses indicated in Figure 5.25 by (G_1,G_2) and which respectively represent the stress relating to the out-of-plane strain and the final thinning. Conversely, a variation of the deterministic optimum pushes it towards the non-feasibility domain, which affects the stability of the process and does not exclude the occurrence of plastic instabilities during the process due to instantaneous uncertainties. As a matter of fact, the mechanical properties, the loading paths applied, as well as the lubrication conditions, can vary from tube to tube, which makes that the prediction of the final state of the part is unpredictable, the deterministic optimum is often risky, and the stability of the process is not guaranteed. The variation region occupied by the reliability optimum is much more stable compared to that occupied by the deterministic optimum.

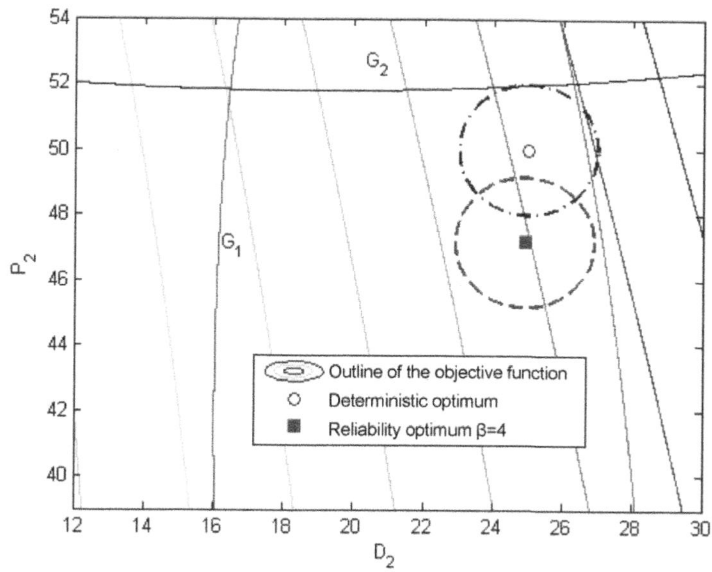

Figure 5.25. *Optimum sensitivity to uncertainties. For a color version of this figure, see www.iste.co.uk/elhami/multi-physics.zip*

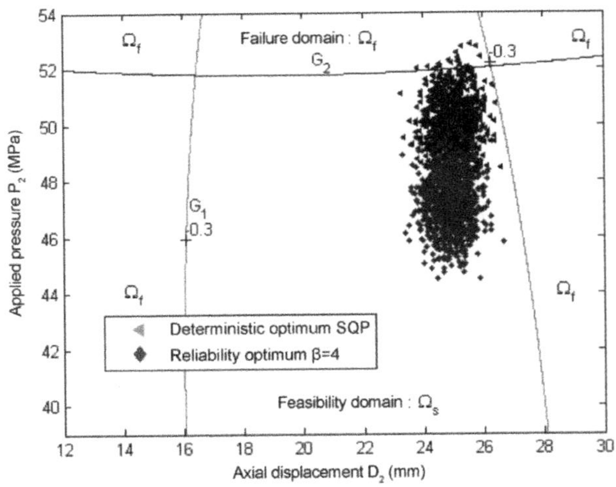

Figure 5.26. *Impact of a perturbation of the optimums with a rate of δ = 2%. For a color version of this figure, see www.iste.co.uk/elhami/multi-physics.zip*

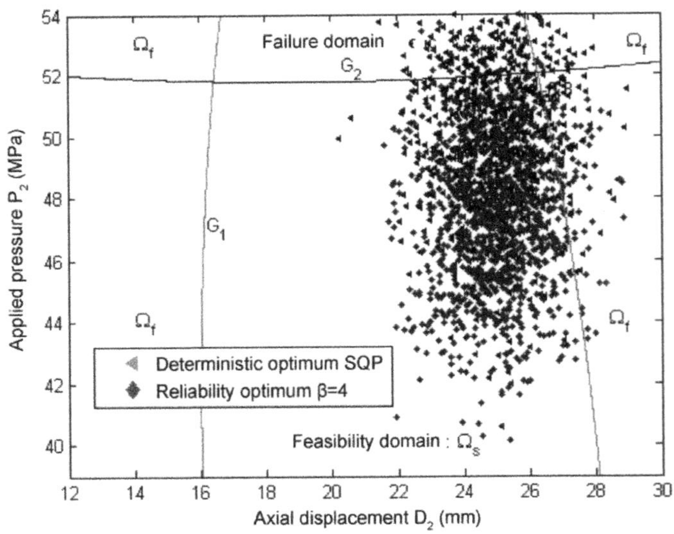

Figure 5.27. *Impact of a perturbation of the optimums with a rate of δ = 5%. For a color version of this figure, see www.iste.co.uk/elhami/multi-physics.zip*

For the same purpose, the deterministic and reliability optimums are perturbed with a defined rate of variation and we look into their variations in the feasibility space. For two different levels of variation that correspond to $\delta = 2\%$ and $\delta = 5\%$, 1,000 random realizations are generated with a Monte Carlo technique and their distributions are analyzed. Figures 5.26 and 5.27 give the distribution of these realizations in space (D_2,P_2) and delineate the feasibility domain denoted by Ω_S from the domains of failure denoted by Ω_f. For the first level of uncertainty, it is noted that only realizations resulting from a perturbation of the deterministic optimum exceed the limits of feasibility. An increase in the level of perturbation affects the deterministic optimum much more than the reliability optimum. Most of the achievements resulting from a perturbation of the reliability optimum are in the admissible domain.

5.5. Deterministic and reliable optimization of a tube with two expansion areas

The second example discussed in this chapter consists of hydroforming a tube with two expansion zones of different heights. The problem is solved with a classical deterministic method and a probabilistic approach in order to compare the two. To simplify the problem, we only consider the loading parameters that are uncertain.

The finite element model is composed of a die with two expansion zones, two punches that ensure the supply of material to the expanded area modeled by rigid bodies. The tube has a diameter of 45 mm and an initial thickness of 1.5 mm. The finite element model is shown in Figure 5.28.

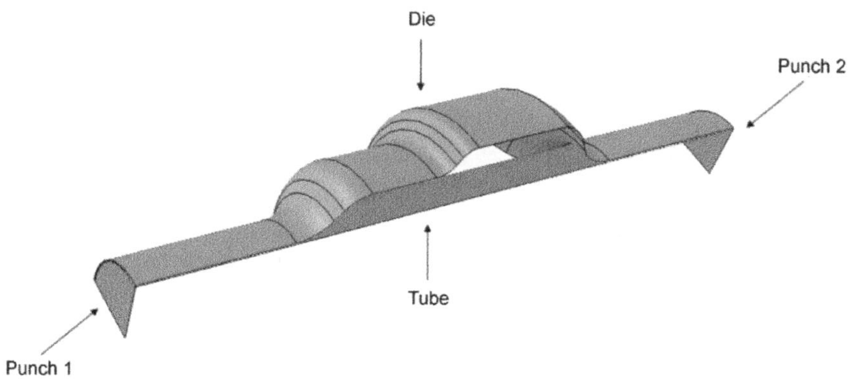

Figure 5.28. *Finite element model. For a color version of this figure, see www.iste.co.uk/elhami/multi-physics.zip*

In order to study the sensitivity of the deterministic and reliable optimum to uncertainties, these two optimums are perturbed to observe their ranges of variation. For the reliability optimum, the one that corresponds to a reliability index $\beta = 4$ is chosen.

In Figure 5.29, it can be clearly seen that the reliability optimum does not exceed the limits imposed by the two stresses noted in Figure 5.29 by G_1 and G_2, which respectively represent the stress relating to the out-of-plane strain and final thinning. Conversely, a variation of the deterministic optimum pushes it towards the non-feasibility domain, which affects the process stability and does not exclude the occurrence of plastic instabilities during the process due to instantaneous uncertainties. As a matter of fact, the mechanical properties, the loading paths applied as well as the lubrication conditions can vary from tube to tube, which makes that the prediction of the final state of the part is unpredictable, leading to a deterministic optimum often risky and does not guarantee the stability of the process. The variation region occupied by the reliability optimum is much more stable compared to that occupied by the deterministic optimum.

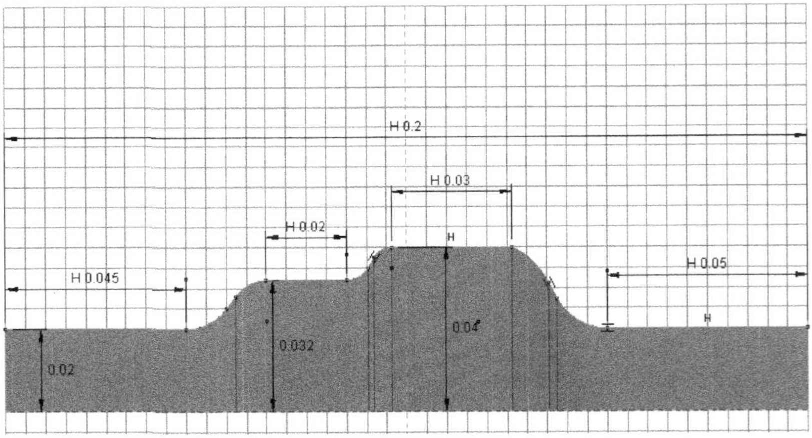

Figure 5.29. *Die dimensions (in meters). For a color version of this figure, see www.iste.co.uk/elhami/multi-physics.zip*

The loading paths used are shown in Figures 5.30 and 5.31. The aim of the problem is to optimize the loading paths in a deterministic and reliable way. The aim is to control the displacements of the two punches and the internal pressure. In this problem, the axial displacement is modeled by the points D_1 for punch 1 and D_2 for punch 2. The internal pressure is controlled by three pressure levels denoted by (P_1, P_2, P_3).

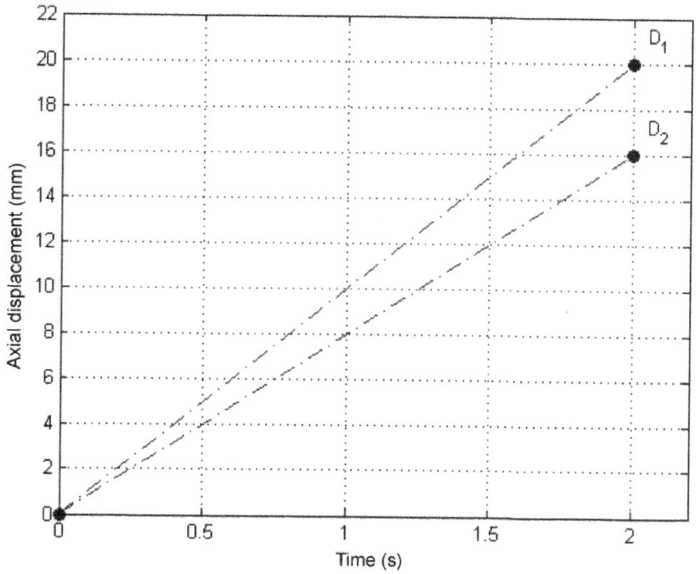

Figure 5.30. *Loading path displacement (mm) – time (s). For a color version of this figure, see www.iste.co.uk/elhami/multi-physics.zip*

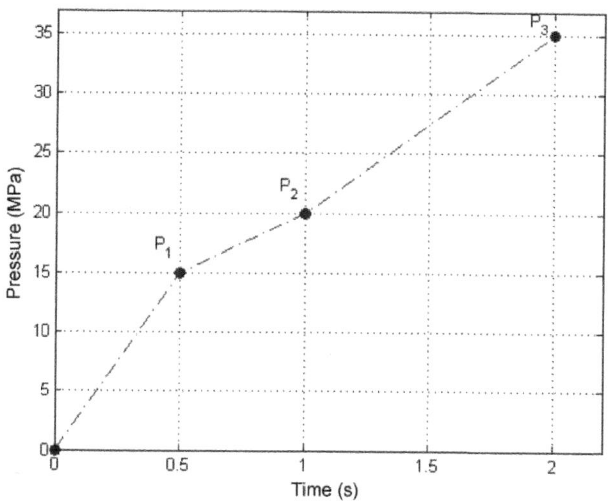

Figure 5.31. *Loading path internal pressure (MPa) – time (s). For a color version of this figure, see www.iste.co.uk/elhami/multi-physics.zip*

Table 5.5 shows these variables with their probabilistic characteristics. These variables are assumed to be statistically independent.

V.a.r.	Average	Var. coef. (%)	Distribution
D_1	20	5	Normal
D_2	16	5	Normal
P_1	15	5	Normal
P_2	20	5	Normal
P_3	35	5	Normal

Table 5.5. *Probabilistic characteristics of loading parameters*

The objective function defined concerns a minimization of the variation of the final thickness (equation [5.4]). It makes it possible to ensure a homogeneous distribution of the final thickness. The first term takes into account excessive thinning which drives the appearance of necking, while the second allows the controlling of the thickening of the tube, which results in controlling the tendency to folding. In other words, it ensures that neither thinning nor thickening would take on excessive values. The objective function is formulated as follows:

$$f(x) = \frac{1}{N}\left(\sum_{i=1}^{N}\left(\frac{h_0 - h_i}{h_0}\right)^2\right)^2 + \frac{1}{N}\left(\sum_{i=1}^{N}\left(\frac{h_i - h_0}{h_0}\right)^2\right)^2 \quad [5.4]$$

where f(x) is the objective function; x is the vector of the optimization parameters; h_i is the thickness of the element i; h_0 is the initial thickness; and N is the total number of elements.

The definition of the constraints concerns the potential plastic instabilities. Folding is taken into account through the objective function which allows us to control tube thickening and also through geometric constraints. For necking, we rely on the material CFL to define a constraint to avoid the occurrence of this instability. Additionally, we rely on the critical main strains as shown in Figure 5.32.

Geometric constraints are also defined that enable the tube to properly assume the shape of the die. These three constraints consist of ensuring that the distance of the nodes at the level of the three radii is minimal. Figure 5.33 describes these types of constraints.

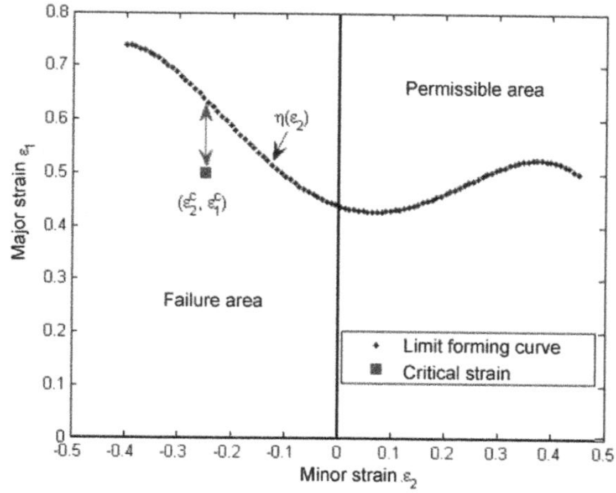

Figure 5.32. *Forming limit curve: necking criterion. For a color version of this figure, see www.iste.co.uk/elhami/multi-physics.zip*

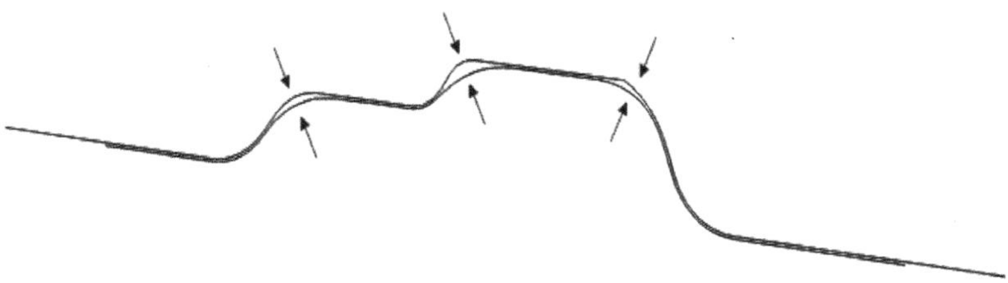

Figure 5.33. *Geometric constraints*

5.5.1. *Deterministic formulation of the optimization problem*

The deterministic optimization problem is written as follows:

$$\begin{cases} \min f(x) \\ g_1(x) = \varepsilon_1^c - \eta\left(\varepsilon_2^c\right) \\ g_i = d_i - d_m; \quad i = \{1,2,3\} \\ x = \{D_1, D_2, P_1, P_2, P_3\} \end{cases} \qquad [5.5]$$

where f represents the objective function; x is the vector of the optimization variables; g_1 is the deterministic constraint on tube necking; g_i are the geometric constraints, with I = (1,2,3); and $\eta(\varepsilon_2)$ is the equation of the FLC.

For the reliability formulation of the problem, the constraint on tube necking is transformed into a probabilistic constraint. The geometric constraints are kept deterministic.

5.5.2. *Reliability formulation of the optimization problem*

The reliability optimization problem is defined as follows:

$$\begin{cases} \min f(\mu_X) \\ P\left[G_1(\mu_X) \leq 0\right] \leq P_{admissible} \\ g_i = d_i - d_{mi} \; ; \; i = \{1,2,3\} \\ \mu_X = \left\{\mu_{D_1}, \mu_{D_2}, \mu_{P_1}, \mu_{P_2}, \mu_{P_3}\right\} \end{cases} \quad [5.6]$$

where $G_1 = -g_1$; P[.] is the probability of occurrence of the plastic instability and μ_x is the average of the uncertain parameters.

5.5.3. *Numerical results*

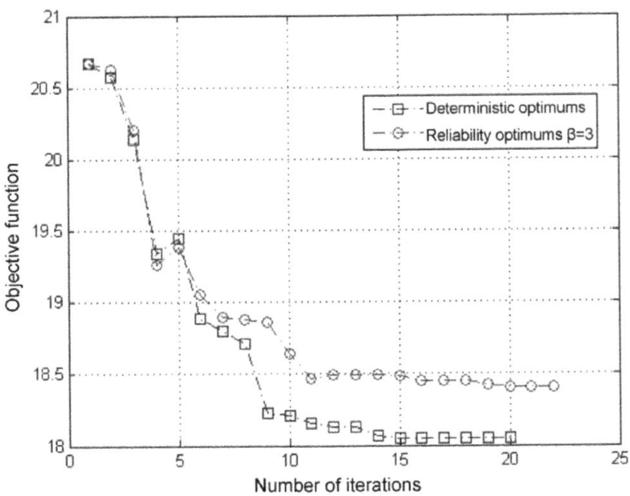

Figure 5.34. *Convergence of the objective function. For a color version of this figure, see www.iste.co.uk/elhami/multi-physics.zip*

The problem is solved with the same approach discussed for the previous example. The convergence for the deterministic and reliability optimums for $\beta = 3$ is summarized in Figure 5.34.

The optimum loading paths are summarized in Figures 5.35–5.36, giving the variation of pressure and axial displacement with respect to time.

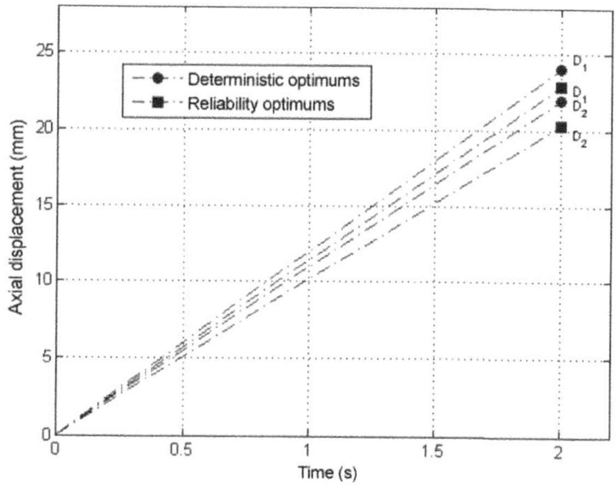

Figure 5.35. *Optimum paths: axial displacement (mm) – time (s). For a color version of this figure, see www.iste.co.uk/elhami/multi-physics.zip*

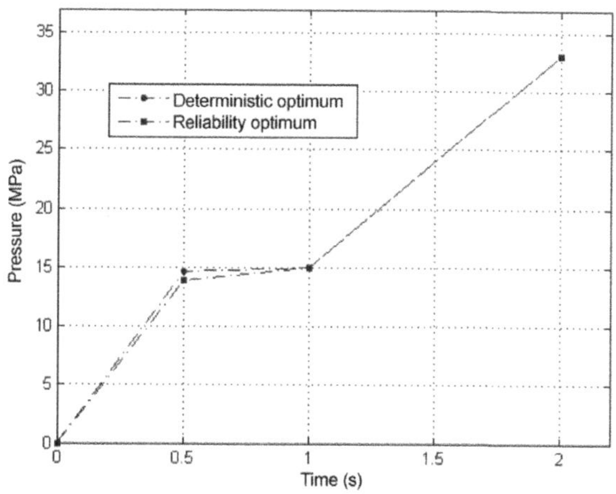

Figure 5.36. *Optimum paths: pressure (MPa) – time (s). For a color version of this figure, see www.iste.co.uk/elhami/multi-physics.zip*

The distribution of the final thickness obtained by the classical deterministic approach and the probabilistic approach is given in Figure 5.37. It is clearly noticeable that the reliability approach allows for a better distribution of thickness and exhibits less thinning, which minimizes the occurrence of necking [ELH 15].

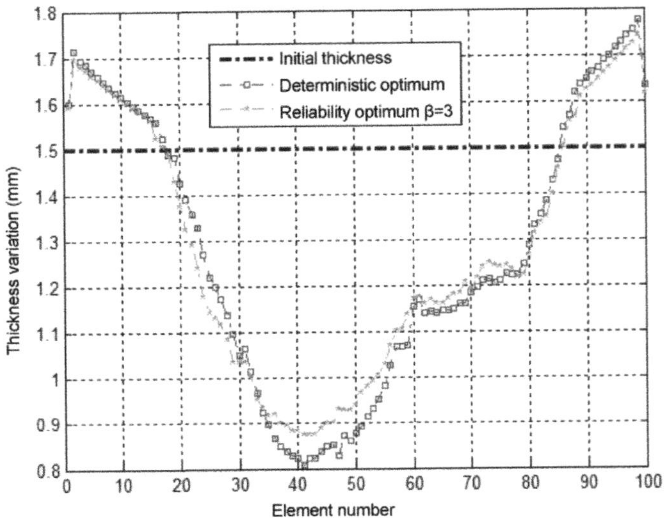

Figure 5.37. *Thickness distribution according to the axial position. For a color version of this figure, see www.iste.co.uk/elhami/multi-physics.zip*

6

Reliability and FSI Optimization

6.1. Introduction

Ensuring the reliability of components exhibiting progressive degradation behavior or acquiring a better knowledge of them in order to optimize their use and reduce their risk of failure – and to do so at the best cost – is a considerable challenge. The theory of reliability uses the estimation of a probability as a measure; whether this probability is acceptable or not is a very complex question. It is clear that a decision involves considering the quality of the information available, the estimated level of risk, the consequences of failure with material and human damage and operational life under risk, knowing that certain risks are imposed on each of them by living conditions, and others are freely consented.

Optimization aims to automatically determine the design that is best with regard to the criteria related to structural performance. The solution found by this iterative process, alternating structural analysis and application of an optimization technique, is called the optimal design. However, any structure optimized without considering reliability can underlay a high risk of failure compared to the probability of failure that the designer wishes it to have. The objective of reliability-based optimization is to propose a design that guarantees the desired level of reliability.

This chapter starts with a presentation of the notions of reliability as well as its objectives and interest in mechanics. We then explain its implementation in detail within the context of reliability optimization problems of the design of structures in vibration and in interaction with a flowing fluid, in order to detect the critical frequency bands that can lead to the destruction (or damage) of the structure to be optimized.

6.2. Reliability optimization in mechanics

As mentioned earlier, the purpose of design is to define the best performance, allowing for a reasonable trade-off to be made between conflicting needs, such as reliability and cost. It is therefore necessary to integrate reliability-based optimization procedures for structural design in dynamics through the use of numerical modeling software. As such, this software is used to analyze numerous possibilities and it should automate the search for the optimal shape, integrating a reliability analysis in order to evaluate the performance of the optimal design.

In the literature, several methods have been used to solve the issues related to reliability-based design optimization (RBDO) in dynamics, and more particularly, we should mention the hybrid method [MOH 05a]. This method was developed by [KHA 02] who merged the two physical and normal spaces. Studies have been carried out on structures with different parameters (stresses, strain and frequencies), and we are now endeavoring to account for resonance phenomena in mechanical structures.

The hybrid method has been adopted for applications in the context of vibration dynamics computations and in order to determine the crucial band in the vicinity of the resonant frequency. Its application showed that this method was executed with great complexity and a very high computation time. To overcome these two drawbacks, a method has been developed called the "safest point (SP) method", which effectively yields the optimal and reliable solution for vibrating structures.

In the following sections, the different methods of RBDO of vibration structures are recalled and we evaluate these methods with a numerical application in the fluid structure interaction to show the efficiency and robustness of the methods presented and the proposed method.

6.2.1. *Deterministic optimization*

During deterministic optimization, the designer introduces safety coefficients in order to increase design safety. These coefficients are used for the most important parameters of the structure, such as loading. The choice of these coefficients is made from experience or from representative tests, or even by calibration from a probabilistic scheme.

The deterministic design optimization process is shown in Figure 6.1. The deterministic optimization loop contains three steps:

1) Description of the geometry using a geometric model (by CAD software).

2) Model analysis by the finite element method.

3) Structure shape optimization by minimizing an objective function such as cost or volume.

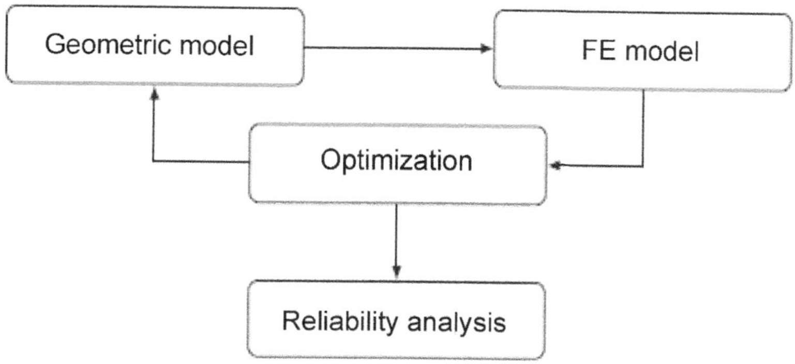

Figure 6.1. *Design optimization process*

Deterministic structural optimization requires numerous structural analyses and failure mode verification. However, in reality, design difficulties come across when describing uncertainties with the variables because optimal solutions may lead to a lower level of reliability, hence the interest in reliability-based optimization.

6.2.2. *Different approaches to RBDO*

Safety factors are calibrated for a wide class of structures and their use in optimization does not guarantee an optimum with appropriate reliability. In addition, the reliability of the optimum obtained is completely overlooked because these coefficients are not directly related to the reliability requirement. This can lead to a lack of robustness of the optimal solution. Reliability-based optimization relies on failure analysis at each iteration in order to determine the roles of uncertainties on the system. Nonetheless, in practical applications, coupling between reliability analysis and optimization procedures leads to very high computational times and low convergence stability.

In the space of random variables, solving the reliability problem involves a large number of calls to the mechanical model, whereas in the space of design variables, the search for the optimal solution modifies the configuration of the structure and thus requires a re-evaluation of its reliability with each iteration. There is therefore a strong motivation to develop efficient techniques to reduce computation time.

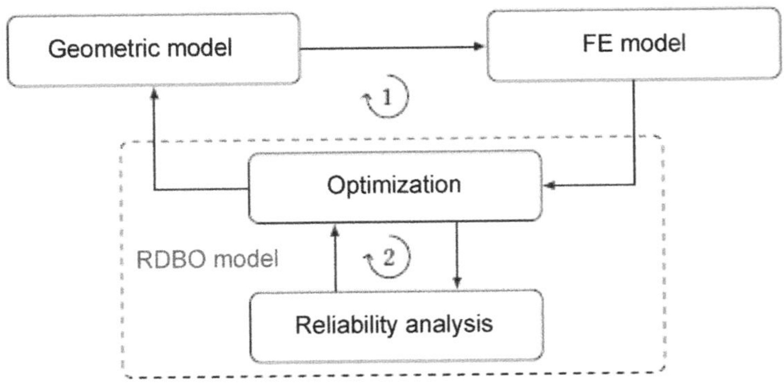

Figure 6.2. *Sequential process of reliability-based design optimization. For a color version of this figure, see www.iste.co.uk/elhami/multi-physics.zip*

The RBDO process (see Figure 6.2) comprises two nested loops:

1) An optimization loop of the objective function such as cost and volume under physical, geometric, functional constraints, etc.

2) A reliability computation loop by an optimization procedure for a given limit state.

To introduce reliability into design optimization, we consider the reliability index as a constraint to be satisfied. In the conventional deterministic optimization process, we introduce safety coefficients for the most significant variables (load, geometry, etc.). Any optimization algorithm can be used to achieve the optimal solution. At the optimal point, reliability analysis is performed to determine the reliability index corresponding to the limit state under consideration. In the classic reliability optimization process, the safety factors are replaced by a reliability limitation, which is directly integrated into the main optimization problem. This requires a re-evaluation of the reliability index and the most likely point of failure at every iteration, in order to meet the target level of reliability.

Reliability-based optimization is a new approach that introduces reliability criteria in the search for the optimal configuration. Its objective is to define the best compromise between cost reduction and ensuring reliability, by integrating uncertainties on the system variables. The optimization procedure must allow the redistribution of roles of uncertainties on the basis of the sensitivity of the design objectives. For this reason, deterministic optimization cannot ensure adequate levels of reliability. A deterministic design can lead to significant economic losses due to failure, while a design that is too reliable is more often than not too expensive. Reliable design optimization aims to achieve a balanced design by reducing the total cost.

The advantage of this approach is that it drives the search for the optimum by controlling the variables that are penalized regardless of their mechanical role. In this regard, the robustness of the system is achieved because the role of uncertainties is reduced during the optimization procedure. From a numerical point of view, reliability optimization approaches require the coupling between mechanical modeling (finite element method), optimization algorithms and structural reliability analysis tools. The classical integration of reliability analysis into optimization methods is carried out in two spaces: the normed space of random variables and the physical space of design variables, which requires a very long computation time.

In the reliability formulation, we make a distinction between two types of variables:

1) The optimization variables x which are deterministic variables to be controlled for optimizing the sizing process; they represent the control parameters of the mechanical system (namely dimensions, materials, loads, etc.) and the probabilistic model (i.e. averages and standard deviations of random variables).

2) Random variables y that represent uncertainties in design variables. Each of these variables is identified by the type of distribution and the associated parameters. These variables can be geometric dimensions, material characteristics or external loads.

6.2.3. *Traditional approach*

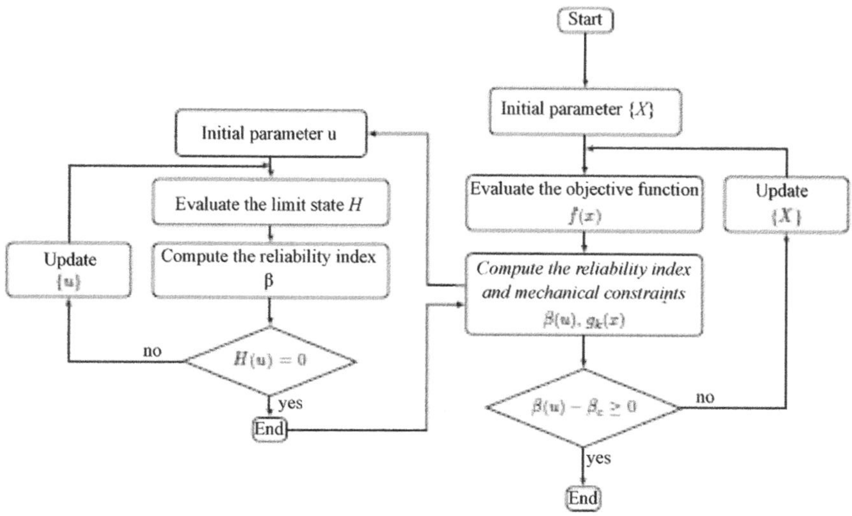

Figure 6.3. *Traditional approach algorithm*

The traditional RBDO consists of first performing the reliability analysis. The resulting solution is used to minimize the objective function under physical, geometric or functional constraints, including the reliability constraint. Convergence is achieved after several iterations alternating between the optimization and reliability procedures. Both of these problems can be solved using any nonlinear programming algorithm such as the SQP method or the penalty method.

In this approach, we obtain the total number of iterations by the cumulation of iterations in the two problems: optimization and reliability, which leads to a very large number of evaluations of the mechanical model [FEN 86, FRA 94]. To overcome this issue, we propose a hybrid reliability optimization formulation.

6.2.4. *Hybrid approach*

This approach consists of integrating the two subproblems into a single problem that is simultaneously solved in the spaces of deterministic and random variables. In this diagram, the reliability problem is included in the formulation of the optimization problem, in order to arrive at a single objective function F(x,y) and, thus, to reduce the overall computational cost. It is expressed as:

$$F(x,y) = f(x) \cdot d_{\beta}(x,y) \quad [6.1]$$

where f(x) and $d_{\beta(x,y)}$ are respectively the objective function and the image of the reliability index in hybrid space (see Figure 6.3).

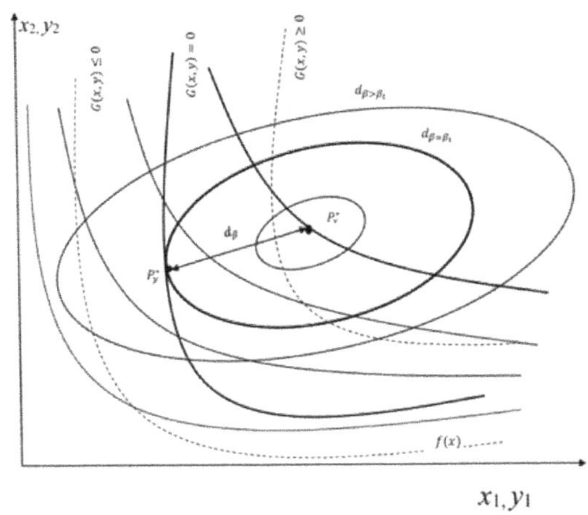

Figure 6.4. *Hybrid design space*

In this space, all of the information regarding the optimization process is grouped together; P^*_y is the optimal solution of the problem and P^*_x is the design point. The problem of the hybrid RBDO is written as:

$$\min_{x,y} \quad F(x,y)$$

$$\text{s.c.} \quad \begin{cases} G(x,y) \leq 0 \\ g_k(x) \leq 0 \\ d_\beta(x,y) \geq \beta_c \end{cases} \qquad [6.2]$$

where $G(x,y) \leq 0$ is the limit state, $g_k(x) \leq 0$ are the deterministic constraints and β_c is the target reliability index.

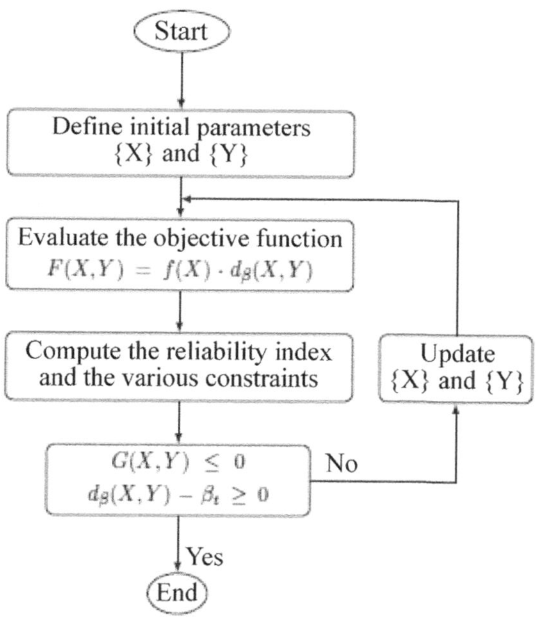

Figure 6.5. *Hybrid method algorithm*

The objective of the new problem is to minimize the function F(x,y) under the active constraint $G(x,y) \leq 0$ which must be strictly satisfied, and the other constraints $g_k(x) \leq 0$ and $d_\beta(x,y) \geq \beta_c$ as available resources to be optimally used. In traditional studies, the reliability index is calculated by an optimization procedure in the standard Gaussian space. However, the hybrid formulation considers both spaces:

deterministic and random. The resolution must be achieved within the hybrid design space (x,y). This procedure leads to a parallel convergence of the two problems.

6.2.5. *Hybrid frequency approach*

The inclusion of reliability optimization procedures has given rise to a new method referred to as hybrid dynamic and adapted to problems involving free-vibration structures [MOH 05b]. The main idea is to search for two design points instead of just one.

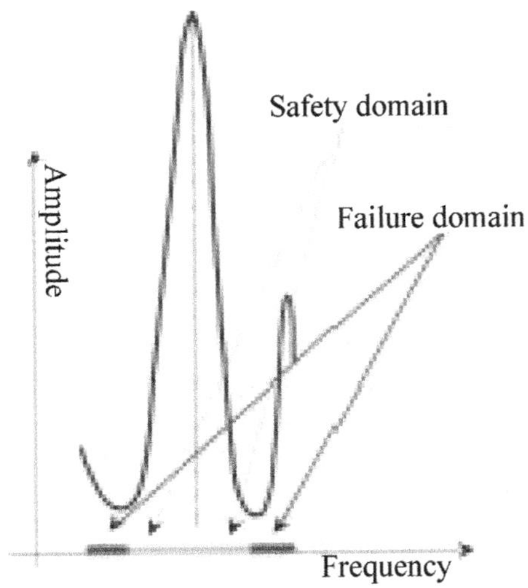

Figure 6.6. *Resonant frequency. For a color version of this figure, see www.iste.co.uk/elhami/multi-physics.zip*

These two points are used to find a lower and upper bound to limit the critical frequency band. This zone coincides with the crucial frequency band in the vicinity of the resonant frequency of the structure to be optimized (see Figure 6.6). These two limits are to be found at each computational iteration.

Optimization problem formulation

The formulation of the vibratory dynamic problem is given by:

$$\min_{x,y} \quad F(x,y) = f(x)\cdot d_{\beta_A}(x,y)\cdot d_{\beta_B}(x,y)$$

$$\text{s.c.} \quad \begin{cases} G(x,y)\le 0 \\ g_k(x)\le 0 \\ d_{\beta_A}(x,y)\ge \beta_t \quad \text{and} \quad d_{\beta_B}(x,y)\ge \beta_t \end{cases} \qquad [6.3]$$

where $d_{\beta A}(x,y)$ represents the image of the reliability index in the hybrid space (HDS: Hybrid Design Space) where $G(x,y) \le 0$ is the limit state, $g_k(x) \le 0$ are the deterministic constraints and β_c is the target reliability index.

Normal distribution

For a normal distribution, the problem contains two design variables that are random variables. In the normal space, the distance d(u) is defined by $d(u) = \sqrt{\Sigma u_i^2}$, where u_i is a normal variable defined by:

$$u_i = \frac{y_i - m_i}{\sigma_i}, \qquad i = 1,\dots,n \qquad [6.4]$$

The average m_i is replaced by the deterministic design variable x_i ($x_i = m_i$); the image of the reliability index in the hybrid space is defined by:

$$d_\beta = \sqrt{\sum_{i=1}^{n}\left(\frac{y_i - x_i}{\sigma_i}\right)^2} \qquad i = 1,\dots,n \qquad [6.5]$$

The reliability levels d_β are represented by ellipses (for a normal distribution).

Log-normal distribution

In many mechanical problems, a random variable cannot reach negative values due to the physical aspects of the problem. In this situation, the log-normal distribution is more appropriate, because it automatically eliminates the possibility of obtaining negative values.

This distribution is directly related to the normal distribution. More specifically, if the random variable X follows the log-normal distribution, ln(X) has a normal distribution that is characterized by two distribution parameters, the average m_i and the standard deviation σ_i. The normal variable u_i is defined by:

$$u_i = \frac{\ln(y_i) - \mu_i}{\xi_i}, \quad i = 1,...n \tag{6.6}$$

where μ_i and ζ_i are the two distribution parameters of the log-normal distribution. They are defined by:

$$\mu_i = \ln\left(\frac{x_i}{\sqrt{1+\gamma_i^2}}\right), i = 1,\ldots,n \qquad \zeta_i = \sqrt{\ln\left(1+\gamma_i^2\right)}, \qquad i = 1,\ldots,n$$

with $\gamma_i = \dfrac{\sigma_i}{x_i}$.

The normal variable u_i can also be expressed by:

$$u_i = \frac{\ln\left(\dfrac{y_i\sqrt{1+(\sigma/x_i)^2}}{x_i}\right)}{\sqrt{\ln(1+(\sigma/x_i)^2)}}, \quad i = 1,\ldots,n \tag{6.7}$$

and the image of the reliability index in the hybrid space is defined by:

$$d_\beta = \sqrt{\sum_{i=1}^{n} \frac{\left(\ln\left(\dfrac{y_i\sqrt{1+(\sigma/x_i)^2}}{x_i}\right)\right)^2}{ln(1+(\sigma/x_i)^2)}}, \qquad i = 1,\ldots,n \tag{6.8}$$

6.3. Safest point (SP) method

The approach that will be introduced has been developed for applications related to problems involving vibration dynamics calculations and for determining the critical frequency bands relating to a structural optimization case; this new method is called the SP method ([KHA 08, RAD 10]). The main idea of this proposed new technique is to look for an optimal and reliable solution for structures in vibration and interaction with a flowing fluid. The solution sought is one that coincides with a point at the same reliability index with respect to the two lower and upper bounds of the critical frequency band $[f_a, f_b]$.

Problem formulation

Consider the equality of the two reliability indices: $\beta_a = \beta_b$ with

$$\beta_a = \sqrt{\sum_{i=1}^{n}\left(u_i^a\right)^2} \quad \text{and} \quad \beta_b = \sqrt{\sum_{i=1}^{n}\left(u_i^b\right)^2} \qquad i=1,\dots,n \tag{6.9}$$

To verify equality [6.9], we propose the equality of each bound. Therefore, it follows that:

$$u_i^a = -u_i^b \quad i=1,\dots,n \tag{6.10}$$

We propose two studies for the normal distribution and the log-normal distribution.

Normal distribution

For a normal distribution, the normalized variable u_i has the following shape:

$$u_i = \frac{y_i - m_{yi}}{\sigma_i}, \qquad i=1,\dots,n \tag{6.11}$$

where the average m_{yi} and the standard deviation σ_i are the two distribution parameters of the normal distribution.

From equation [6.10], we have:

$$\frac{y_i^a - m_i}{\sigma_i} = -\frac{y_i^b - m_i}{\sigma_i}, \quad \text{or} \quad \frac{y_i^a - x_i}{\sigma_i} = -\frac{y_i^b - x_i}{\sigma_i}, \quad i=1,\dots,n \tag{6.12}$$

The SP satisfying the equality of the two reliability indices ($\beta_a = \beta_b$) is located in the middle of the interval $[f_a, f_b]$, and corresponds to the frequency f_n:

$$m_i = x_i = \frac{y_i^a + y_i^b}{2}, \quad i=1,\dots,n \tag{6.13}$$

Log-normal distribution

For a log-normal distribution, the normal variable u_i is defined by:

$$u_i = \frac{\ln(y_i) - \mu_i}{\xi_i}, \quad i = 1,\ldots,n \qquad [6.14]$$

where μ_i and ζ_i are the two distribution parameters of the log-normal distribution and which are defined by:

$$\mu_i = \ln\left(\frac{x_i}{\sqrt{1+\gamma_i^2}}\right), \quad i = 1,\ldots,n$$
$$\zeta_i = \sqrt{\ln(1+\gamma_i^2)}, \quad i = 1,\ldots,n \qquad [6.15]$$

with $\gamma_i = \dfrac{\sigma_i}{x_i}$.

The normal variable u_i can also be expressed by:

$$u_i = \frac{\ln\left(\dfrac{y_i\sqrt{1+(\sigma/x_i)^2}}{x_i}\right)}{\sqrt{\ln(1+(\sigma/x_i)^2)}}, \quad i = 1,\ldots,n \qquad [6.16]$$

From equation [6.10], it follows that:

$$\frac{\ln(y_i^a) - \mu_i}{\xi_i} = -\frac{\ln(y_i^b) - \mu_i}{\zeta_i}, \qquad i = 1,\ldots,n \qquad [6.17]$$

where $\mu_i = \ln\left(\dfrac{x_i}{\sqrt{1+\gamma_i^2}}\right) = \dfrac{\ln(y_i^a) + \ln(y_i^b)}{2}$.

In this case, the SP corresponding to the frequency f_n and located in the interval $[f_a,f_b]$ is given by:

$$m_i = x_i = \sqrt{1+\gamma_i^2}\,\exp\left(\frac{\ln(y_i^a \cdot y_i^b)}{2}\right), \qquad i = 1,\ldots,n \qquad [6.18]$$

SP approach implementation

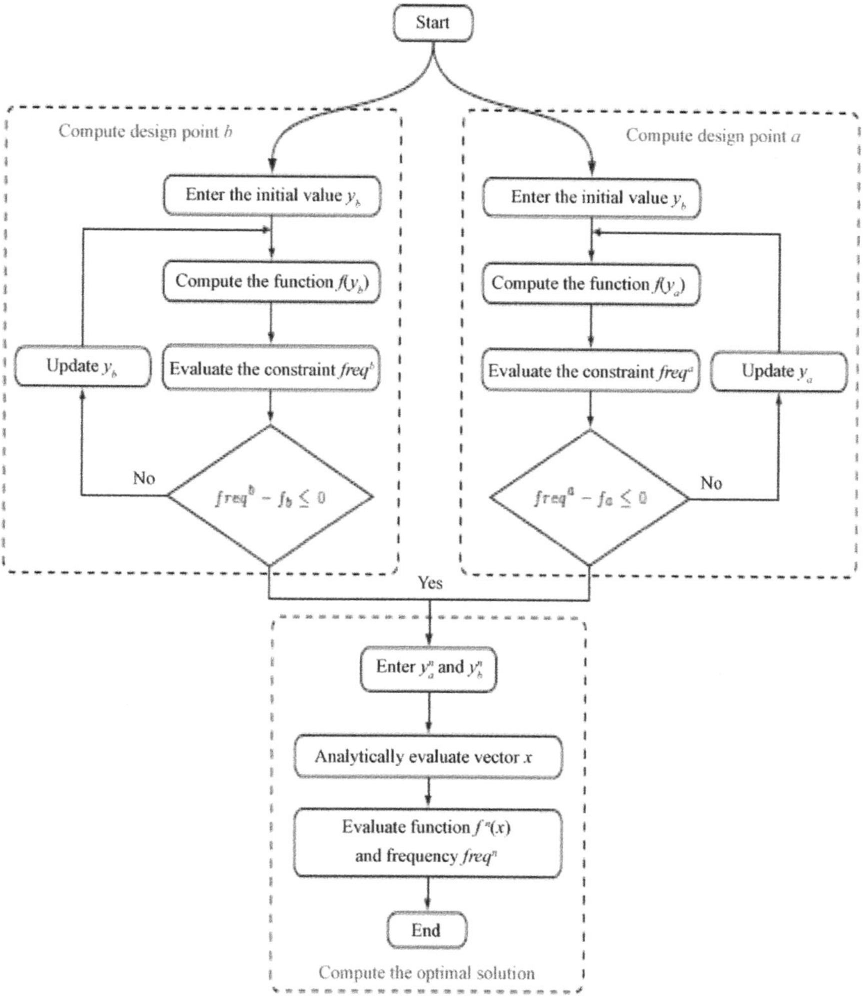

Figure 6.7. *SP approach implementation. For a color version of this figure, see www.iste.co.uk/elhami/multi-physics.zip*

The SP approach can be expressed following three steps:

– *Step 1*. Determine the first design point (or the most likely point of failure): the optimization problem carried out in the physical space consists of minimizing the objective function of the first model under the constraint related to the frequency f_a

(f_a is the lower bound of the critical frequency band). The resulting solution is considered the most likely point of failure and is referred to as design point A.

– *Step 2*. Determine the second design point: the second optimization problem consists of minimizing the objective function of the second model under the constraint related to the frequency f_b (f_b is the upper bound of the critical frequency band). The resulting solution is considered the most likely point of failure and is referred to as design point B.

– *Step 3*. Calculate the optimal solution: in the last step, we determine the optimal design of the structure by a simple calculation. This solution coincides with a point corresponding to the resonant frequency f_n, and located at the same reliability index with respect to the two lower and upper bounds of the critical frequency band $[f_a,f_b]$.

6.4. Numerical simulation

6.4.1. *Reliability calculation for an aircraft wing*

6.4.1.1. *Probabilistic study*

In this application, we use the previous results from the example of the aircraft wing interacting with a flowing fluid in the problem addressed with the finite element method. Naturally, the results of a deterministic analysis are only as good as the assumptions and input values used for that analysis.

The validity of these results depends on the values taken for the component under real conditions. In practice, any model analysis is subject to uncertainty.

It is neither physically nor financially possible to completely eliminate the dispersion of input parameters. Its reduction is usually associated with higher costs, either through better and more accurate manufacturing methods or improved quality control.

A probabilistic study is used to determine the impact of one or more variables on the analysis results; given the complexity of the problem and for the sake of simplicity, we have chosen in our case to take into account only the sources of uncertainties related to the properties of materials, and not to take into account the uncertainties related to the other structural elements (geometry, boundary conditions, etc.).

Parameters	Averages	Standard deviation	Distribution
Young's modulus (Pa)	7.1*e*10	0.355*e*10	Gaussian
Density Kg.m^{-3}	2770	83.1	Gaussian

Table 6.1. *Random variables and statistical distributions*

In this context, stochastic calculus was performed using the probabilistic design system of the ANSYS code. This tool is based on computations with a Monte Carlo simulation (MC) for 100 samples and the response surface method (RSM) for nine samples. Table 6.2 shows the averages and standard deviations of the natural frequencies of the airflow around the wing.

Modes	Deterministic case	MC	RSM
F1	12.630	12.540	12.540
F2	19.182	19.023	19.022
F3	51.426	50.812	50.821
F4	72.938	72.290	72.312
F5	89.837	88.882	88.900
F6	126.02	125.351	124.28

Table 6.2. *Natural frequencies of the wing*

6.4.1.2. *Reliability study*

In this numerical study, the reliability-based methods proposed and used to calculate the probability of failure are the FORM and SORM methods taking into account the uncertainties related to structural properties [RAB 10]. The numerical calculation is performed using a MATLAB coupling code to evaluate the reliability of the structure and ANSYS for the calculation of the implicit limit-state function G which is defined with respect to the first natural frequency F_1 of the coupled system and limited by F_0 such that:

$$G(E,_{\varrho}) = F_1 - F_0 \text{ with } F_0 = 11.8 \text{ Hz} \quad [6.19]$$

Table 6.3 includes the design parameters and their statistical moments considered for the coupled wing and the results obtained by the FORM and SORM methods [RAB 17].

Parameters	FORM	SORM
Young's modulus (Pa)	7.793e10	7.7927e10
Density (Kg/m^3)	3007.1	3007.2
Reliability index β	3.4575	3.5126
Probability P_f	0.00027258	0.0002219
Reliability	99.973	99.978

Table 6.3. *Random variables and their statistical moments*

6.4.2. *RBDO application to the aircraft wing*

The ONERA M6 wing was designed by Bernard Monnerie and his colleagues in 1972, as part of the cooperation within the AGARD (Advisory Group for Aerospace Research and Development), to serve as an experimental support for three-dimensional flow studies at transonic speeds and high Reynolds numbers and for the validation of numerical flow simulations (conditions representative of real military and civilian aircraft flights). It has a symmetrical profile, and its dimensions are shown in Figure 6.8.

Figure 6.8. *The M6 wing in ONERA's S2MA wind tunnel*

The results of these tests were recorded in 1979 in an AGARD report by Volker Schmitt and François Charpin [SCH 74]. The ONERA-M6 wing result database has been used hundreds of times to validate CFD (Computational Fluid Dynamics) software and is still used worldwide. It is one of the most popular test cases, particularly suitable for understanding and evaluating models of laminar–turbulent transition, shock–boundary layer interaction, detachment, as phenomena characteristic of what occurs on wings when approaching the speed of sound.

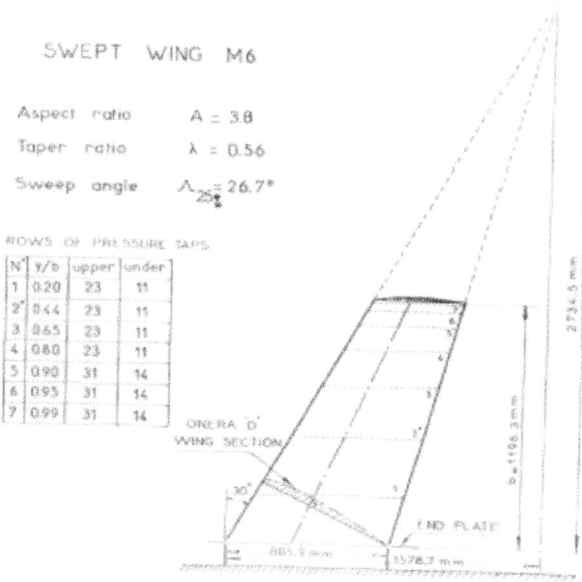

Figure 6.9. *ONERA-M6 wing dimensions*

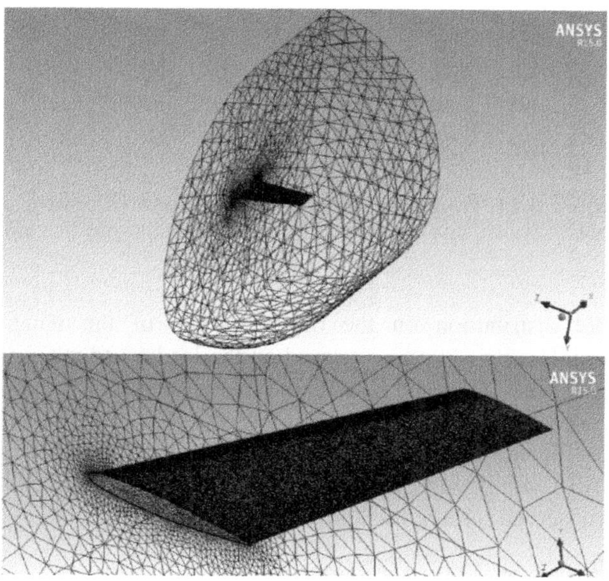

Figure 6.10. *Fluid mesh. For a color version of this figure, see www.iste.co.uk/elhami/multi-physics.zip*

An unstructured hybrid mesh was generated with 375 263 cells for CFD computation as shown in Figure 6.10. It also contains a combination of tetrahedra, pyramids and prisms in the boundary layer region.

6.4.2.1. *Aerodynamic analysis of the ONERA M6 wing*

The aerodynamic analysis was performed by the ANSYS/Fluent solver at a Mach number (M) of 0.8395 with an attack angle (α) of 3.06° and a Reynolds number (Re) of 11.72E6.

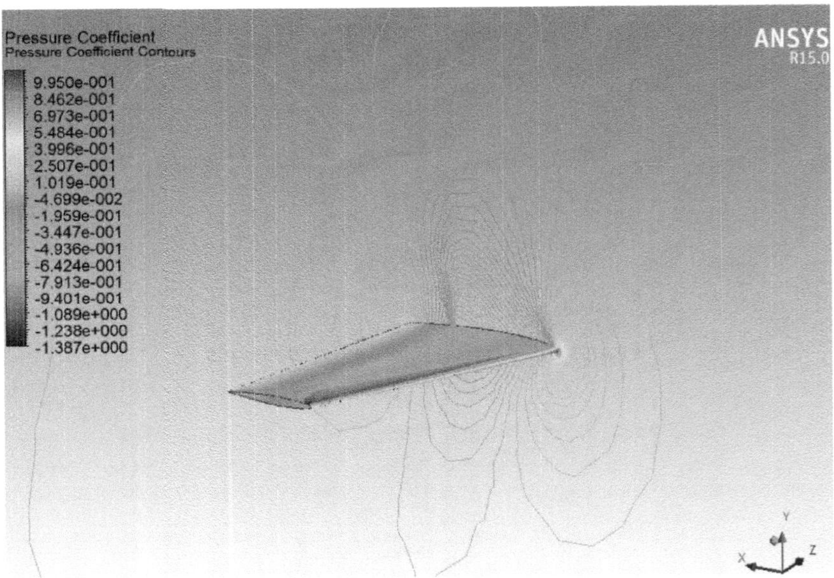

Figure 6.11. *Wing pressure distribution. For a color version of this figure, see www.iste.co.uk/elhami/multi-physics.zip*

The pressure distribution on the upper surface of the wing is shown in Figure 6.11. A violent impact was observed on the leading edge near the fixed end, and this obtained shock weakened close to the free wing end.

The drag coefficient and the lift coefficient are compared (see Figure 6.12) to those of the NASA simulation computed by the WIND code; a good agreement of the results was observed for a refined mesh.

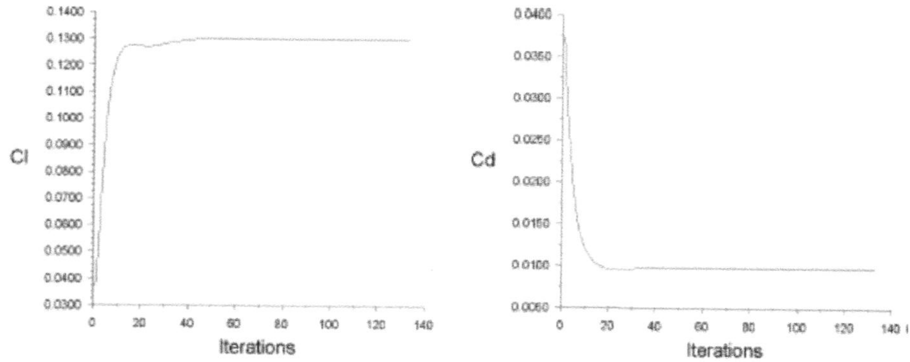

Figure 6.12. *Evolution of the drag and lift coefficient of the ONERA M6 wing according to the number of iterations*

Coefficients	NASA CFD	Original mesh	Refined mesh
Cl	0.1410	0.1279	0.134
Cd	0.0088	0.0111	0.0096

Table 6.4. *Comparison of drag (Cd) and lift (Cl) coefficients*

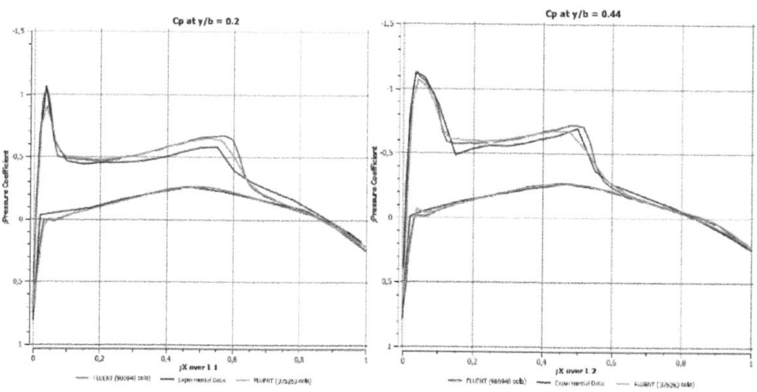

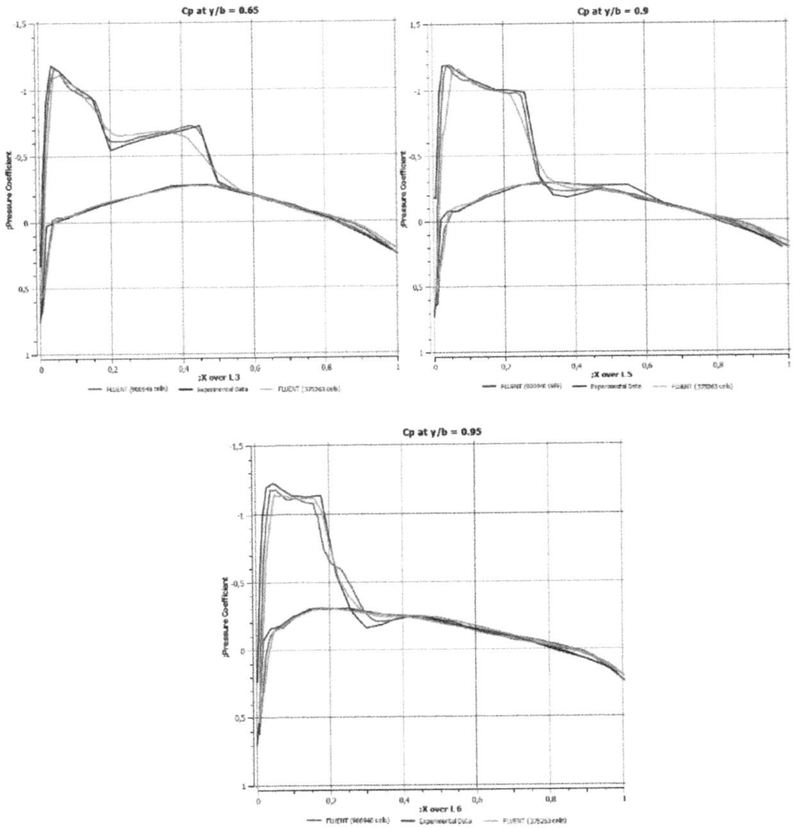

Figure 6.13. *Distribution of the pressure coefficient on the contour of the M6 wing cutting planes according to the abscissa at the intrados and extrados. For a color version of this figure, see www.iste.co.uk/elhami/multi-physics.zip*

Our simulation results are validated by comparing the results found by ANSYS/Fluent with the experimental data for the Onera M6 wing. We plot the pressure coefficient with different contours of the wing cutting planes. A comparison of the results at locations y/b = 0.2, 0.44, 0.65, 0.9 and 0.95 is made between the simulated results and the experimental data [SCH 84] for a first coarse mesh and our more refined mesh, and a good agreement was observed between the results found (see Figure 6.13).

6.4.2.2. *Aeroelastic analysis of the ONERA M6 wing*

In this section, the deformation due to the aerodynamic load of the wing is considered by performing a one-way fluid coupling structure for a static structure.

After developing the aerodynamic load on the wing using ANSYS/Fluent, the pressures on the contact areas are input as pressure loads in ANSYS/Mechanical to determine the stresses and strain on the wing.

The configuration of the structure under consideration is made of an aluminum alloy whose material properties are: the Young modulus (E) is 71 GPa, the Poisson ratio υ is 0.32 and the matter density is 2,770 Kg/m^3.

The calculated wing displacement in the deformed and non-deformed states is shown in Figure 6.14. It can be clearly seen that the calculated fluid pressure has been successfully transferred to deform the CSD volume mesh.

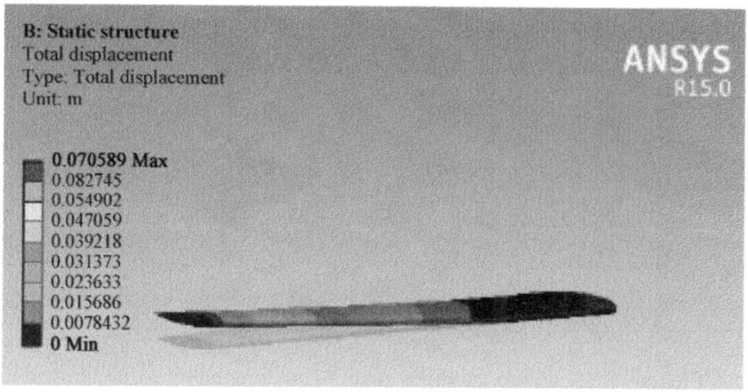

Figure 6.14. *Total wing displacement. For a color version of this figure, see www.iste.co.uk/elhami/multi-physics.zip*

Prestressed modal analysis

The eigenvalue and eigenvector problem is solved with the following equation:

$$[K]\phi_i = \lambda_i [M]\phi_i \qquad [6.20]$$

where [K] is the structure stiffness matrix, φ_i is the eigenvector, λ is the eigenvalue and [M] is the mass structure matrix.

For prestressed modal analyses, the matrix [K] includes the rigid stress matrix. The results of the modal analysis are presented in Figure 6.15. The result comprises the first four mode forms with its natural frequency values.

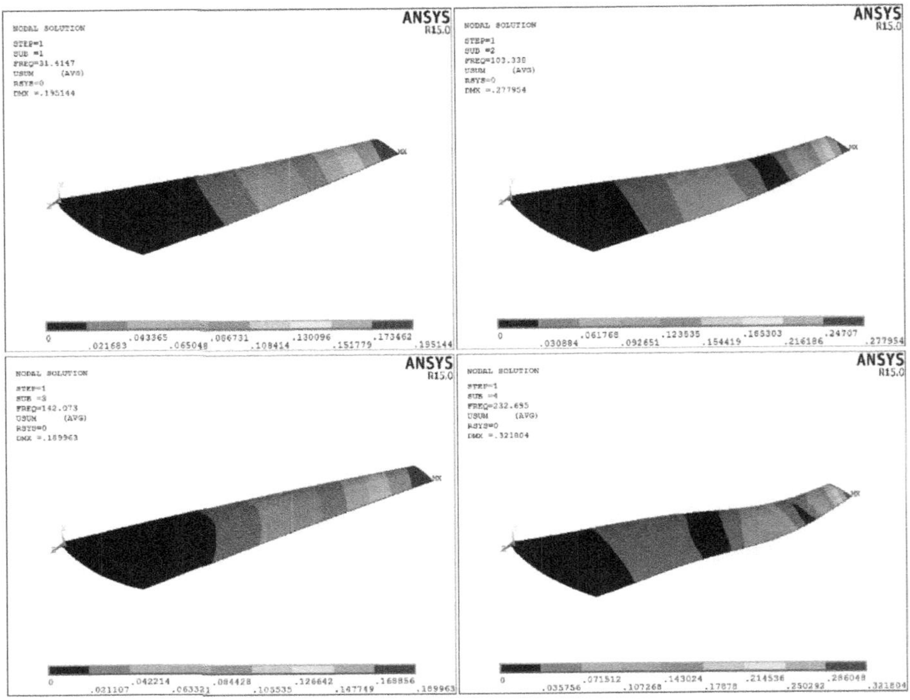

Figure 6.15. *Wing eigenmodes. For a color version of this figure, see www.iste.co.uk/elhami/multi-physics.zip*

6.4.2.3. *Reliable design optimization*

In the following, a comparative study of the hybrid RBDO method and the new SP method has been conducted by way of this application. The advantages of the SP method are studied using two probability distributions, namely the normal distribution and the log-normal distribution to minimize the cross-section of the aircraft wing under a frequency constraint.

The dimensions of the section are A–D, shown in Figure 6.16 and Table 6.5.

Variables	A	B	C	D
Dimensions (m)	0.054	0.092	0.096	0.044

Table 6.5. *Wing parameterization*

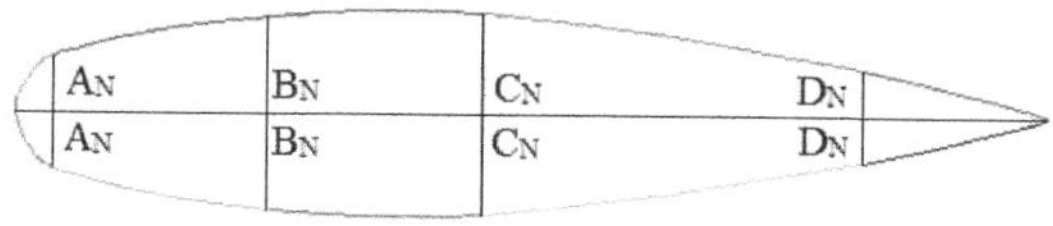

Figure 6.16. *Section dimensions. For a color version of this figure, see www.iste.co.uk/elhami/multi-physics.zip*

The main objective is to find the natural frequency, for a given interval [25,35], located at the safest position of this interval. Thereby, $F_a = 25$ Hz, $F_B = 35$ Hz and $f_n = ?$ Hz, where f_n must verify the equality of the reliability indices: $\beta_A = \beta_b$. The three models can be addressed: the first structure must be optimized under the constraint of the first value of the given frequency F_a; the second must be optimized under the constraint of the second frequency F_B, and the third structure must be optimized under the constraint of a frequency f_n, which verifies the equality of the reliability indices with respect to the two sides of the given interval (see Figure 6.17).

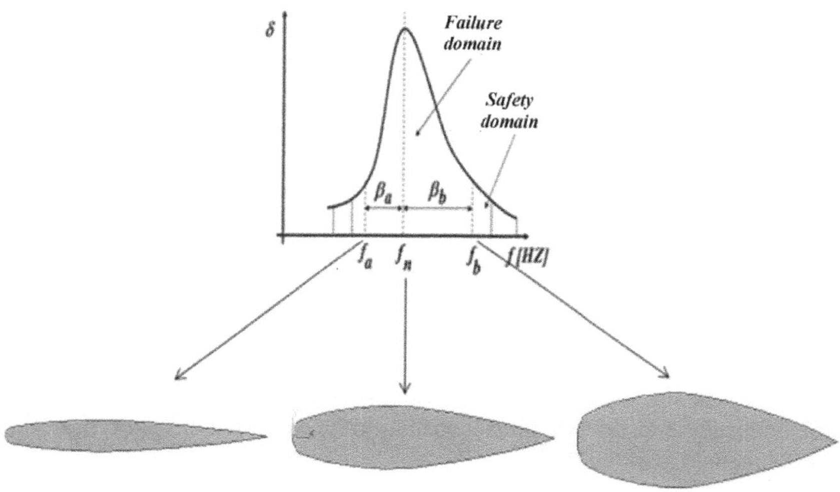

Figure 6.17. *Sections of the three models. For a color version of this figure, see www.iste.co.uk/elhami/multi-physics.zip*

Table 6.6 lists the parameters used for the three models.

Variables		Parameters	Minimum value (MIN)	Maximum value (MAX)
Design values	F_N	A_N	0.027	0.081
		B_N	0.046	0.138
		C_N	0.048	0.144
		D_N	0.022	0.066
	F_A	A_1	0.027	0.054
		B_1	0.046	0.092
		C_1	0.048	0.096
		D_1	0.022	0.044
	F_B	A_2	0.054	0.081
		B_2	0.092	0.138
		C_2	0.096	0.144
		D_2	0.044	0.066
State variables		β_a	3	5
		β_b	3	5
		F_A	25	26
		F_B	34	35
		F_N	25	35
		DIF	-0.01	0.01

Table 6.6. *Parameters for all three models*

Hybrid method

We minimize the volume of the aircraft wing under the constraint of the first natural frequency and the reliability constraint, the formulation of the problem by the hybrid method is of the form:

$$\min_{A_1,\ldots,m_A,\ldots,A_2,\ldots} \quad \text{Volume}\left(\mathrm{m_A},\mathrm{m_B},\mathrm{m_C},\mathrm{m_D}\right)\cdot d_{\beta_1}\left(\mathrm{A_1},\mathrm{B_1},\mathrm{C_1},\mathrm{D_1},\mathrm{m_A},\mathrm{m_B},\mathrm{m_C},\mathrm{m_D}\right)\cdot d_{\beta_2}\left(\mathrm{A_2},\mathrm{B_2},\mathrm{C_2},\mathrm{D_2},\mathrm{m_A},\mathrm{m_B},\mathrm{m_C},\mathrm{m_D}\right)$$

$$s.c. \quad \begin{cases} f_a(A_1,B_1,C_1,D_1) - f_a \le 0, \qquad \mathrm{f_b}(\mathrm{A_2},\mathrm{B_2},\mathrm{C_2},\mathrm{D_2})\text{-}\mathrm{f_b} \le 0 \\ d_{\beta_1}\left(A_1,B_1,C_1,D_1,m_A,m_B,m_C,m_D\right) - d_{\beta_2}\left(A_2,B_2,C_2,D_2,m_A,m_B,m_C,m_D\right) \le 0 \\ u_i^a + u_i^b = 0, \qquad i = 1,\ldots,4 \end{cases}$$

where A–D are grouped in the random vector {y} having as averages the nominal values m_A, m_B, m_C and m_D grouped in a vector denoted by {x} with a standard deviation of 0.1.

The following notations are used:

– volume (m_A, m_B, m_C, m_D): the wing volume;

– $f_a(A_1,B_1,C_1,D_1)$: the lower bound of the critical frequency band;

– $f_b(A_2,B_2,C_2,D_2)$: the upper bound of the critical frequency band;

– $d_{\beta i}(.)$: represents the image of the reliability index in the hybrid space.

SP method

In this part, the problem formulation is simplified, and we now have two optimization problems:

– The first consists of minimizing the objective function of the first model under the constraint related to the frequency f_a:

$$\begin{array}{ll} \min\limits_{A_1,\ldots} & \mathrm{Vol}_a\left(A_1,B_1,C_1,D_1\right) \\ s.c. & f_{\max}\left(A_1,B_1,C_1,D_1\right)-f_a=0 \end{array} \qquad [6.21]$$

– The second consists of minimizing the objective function of the second model under the constraint related to the frequency f_b:

$$\begin{array}{ll} \min\limits_{A_2,\ldots} & \mathrm{Vol}_b\left(A_2,B_2,C_2,D_2\right) \\ s.c. & f_{\max}\left(A_2,B_2,C_2,D_2\right)-f_b=0 \end{array} \qquad [6.22]$$

– After the optimization problems, we calculate the coordinates of the optimal point that corresponds to the third model under the frequency f_n. The average solution of the SP coincides with a point that is located in the middle of the interval $[f_a,f_b]$.

For a normal distribution:

$$m_i = x_i = \frac{y_i^a + y_i^b}{2}, \qquad i=1,\ldots,n \qquad [6.23]$$

For a log-normal distribution:

$$m_i = x_i = \sqrt{1+\gamma_i^2}\exp\left(\frac{\ln(y_i^a \cdot y_i^b)}{2}\right), \qquad i = 1,\ldots,n \qquad [6.24]$$

6.4.2.4. *Results and discussion*

Tables 6.7 and 6.8 show the results of the RBDO with both hybrid and SP methods. The results obtained using two different distribution distributions are presented, in order to show the effectiveness of the proposed new method [ELH 13].

For a normal distribution

Table 6.7 presents the results of the RBDO with the two methods, hybrid and SP using the normal distribution.

Variables	Parameters	Initial	Optimal design with SP	Optimal design with HM
F_N	A_N	0.054	0.040826	0.057710
	B_N	0.092	0.069145	0.084087
	C_N	0.096	0.095406	0.093940
	D_N	0.044	0.037493	0.037192
F_A	A_1	0.027	0.027285	0.045907
	B_1	0.046	0.046169	0.071424
	C_1	0.048	0.082863	0.080665
	D_1	0.022	0.028582	0.032862
F_B	A_2	0.081	0.054367	0.067262
	B_2	0.138	0.092121	0.094616
	C_2	0.144	0.10795	0.10657
	D_2	0.066	0.046404	0.045188
State variables	F_A	23.746	24.998	25.810
	F_B	46.788	34.003	34.069
	F_N	31.414	29.102	30.057
	DIF = $\beta_a - \beta_b$	–	1.55e-15	0.013101
	MASS	411.310	359.855	387.794
	Time (s)	-	1100	15270

Table 6.7. *Normal distribution results*

For normal distributions, our goal is to provide the best trade-off between design cost and safety; this objective is easily achieved by using the new SP method, which gives a better solution compared to the hybrid method. The optimal design point obtained by the SP method exactly satisfies the target reliability level $\beta_t = 3$; on the other hand, with the hybrid method, we obtain a reliability level $\beta_t = 3.0153$ (i.e. within the small tolerances of 0.51%).

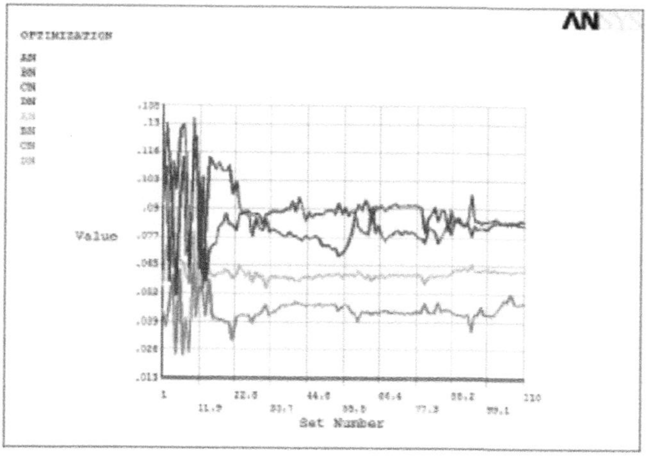

Figure 6.18. *Variation of design variables according to iterations. For a color version of this figure, see www.iste.co.uk/elhami/multi-physics.zip*

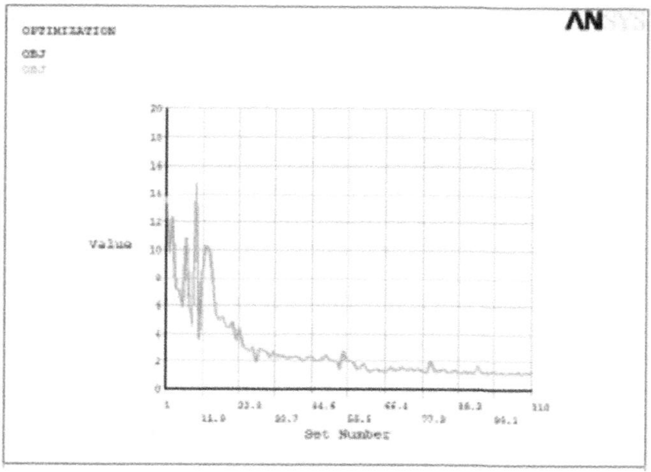

Figure 6.19. *Variation of the objective function according to iterations. For a color version of this figure, see www.iste.co.uk/elhami/multi-physics.zip*

The volume of the aircraft wing is slightly improved with the hybrid method, compared to that obtained by the SP method; however, the numerical cost of the hybrid method is much higher than that of the SP method. Figures 6.18 and 6.19 show the variation in objective function and design variables according to the iterations for the hybrid method [RAB 17].

Log-normal distribution

Table 6.8 presents the results of the RBDO with both the hybrid method and the SP method using the log-normal distribution. Figures 6.20 and 6.21 show the variation in the objective function and design variables according to iterations for the hybrid method.

Variables	Parameters	Initial	Optimal design with SP	Optimal design with HM
F_N	A_N	0.054	0.038582	0.054411
	B_N	0.092	0.070540	0.087850
	C_N	0.096	0.095833	0.079661
	D_N	0.044	0.035276	0.044905
F_A	A_1	0.027	0.027285	0.037127
	B_1	0.046	0.046169	0.087578
	C_1	0.048	0.082863	0.071878
	D_1	0.022	0.028582	0.037519
F_B	A_2	0.081	0.054367	0.071352
	B_2	0.138	0.092121	0.092247
	C_2	0.144	0.10795	0.10706
	D_2	0.066	0.046404	0.046619
State variables	F_A	23.746	24.998	25.95
	F_B	46.788	34.003	34.11
	F_N	31.414	29.102	28
	DIF = $\beta_a - \beta_b$	–	1.55e-15	0.015748
	MASS	411.310	356.410	387.78
	Time (s)	-	1100	16510

Table 6.8. *Log-normal distribution results*

For log-normal distributions, both hybrid and SP methods exactly satisfy the target reliability level $\beta_t = 3$, while obtaining roughly equal optimal volumes, but with a higher computation time using the hybrid method.

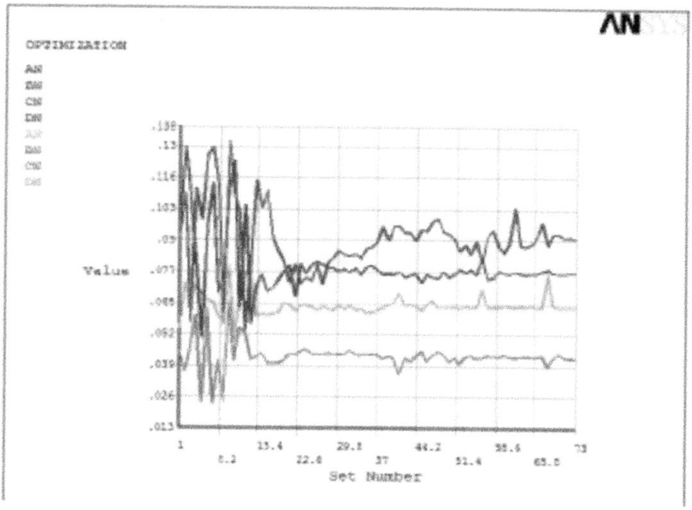

Figure 6.20. *Variation of design variables according to iterations. For a color version of this figure, see www.iste.co.uk/elhami/multi-physics.zip*

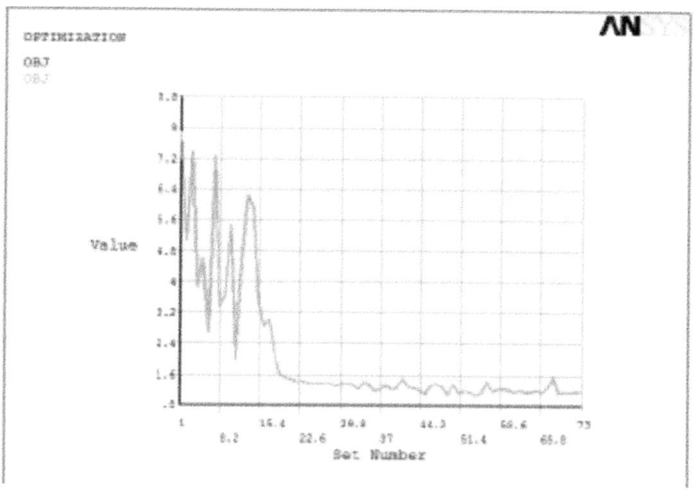

Figure 6.21. *Variation of the objective function according to iterations. For a color version of this figure, see www.iste.co.uk/elhami/multi-physics.zip*

7

Reliability-based Optimization for Dental Implants

7.1. Introduction

Orthodontic dentures can be used to replace teeth, reconstruct them, or in the event of fractures. The dental implant is usually composed of two parts: an artificial root that is fixed in the bone and comes out of the gum, whose function is to support a prosthetic tooth (or element). [PRO 01] have studied the design of dental restorations. More recently, [COU 06] have studied shape optimization for cavity preparation. This study consists of using shape optimization technology to design alternative shapes for dental implants such that the constraint distribution along the bone-implant interface can be optimized.

In [HED 11], the optimization of the dental implant is done using the concept of functionally classified materials. It is widely known that the numerical study of such a dental implant without accounting for the uncertainties of different parameters has been the focus of much attention. However, the parameters of the system may suffer from a lack of accuracy. In this context, the variability of system parameters is a complex field that consists of identifying and modeling the sources of uncertainty, the propagation of uncertainty and post-processing in order to measure the influence of this variability on the system behavior [DAM 17]. The main methods that include uncertainties are probabilistic with uncertain parameters that are modeled as random variables by their probability density functions [BRU 07]. The probabilistic approaches include the Monte Carlo (MC) method [HAM 64], which is widely used in this field. However, it takes a long time to get accurate results. Consequently, the MC method is used as a method for validating the effectiveness of other approaches [GUE 16]. The spectral stochastic approach is also a solution for stochastic problems. Generalized polynomial chaos (GPC) is considered to be one of the spectral stochastic methods. This approach is more effective than other methods such as the MC method. In various studies, the CPG method is employed and compared to other uncertainty methods [SAH 13]. The originality of this study lies in increasing the robustness of the dental implant, which works in the presence of uncertainties. Reliability-based design optimization (RBDO) is coupled with the stochastic approach, in order to obtain the best design of the structure that guarantees the robustness and the required level of reliability.

This chapter is organized into six sections, including the introduction and the conclusion. Section 7.2 presents the description of the stochastic approach such as the MC method and the polynomial chaos method (GPC). The chapter then moves on to discuss RBDO methods and outlines the numerical results in detail.

7.2. Stochastic approach

There are different existing techniques for modeling uncertainty propagation. These methods are classified into three categories: simulation methods, perturbation methods and spectral methods. The MC method is considered a reference method for the calculation on systems with uncertain parameters. The major disadvantage of this method is the very large amount of computation, which complicates its use.

The perturbation method is based on a Taylor series expansion of the response around its mean. The main problem with this method derives from the conditions that ensure the convergence of these series. The variables must have a low dispersion. The method adopted to account for uncertainty is the GPC method. The main idea underlying the polynomial chaos method invented by Norbert Wiener in 1938 [WIE 38] is to establish a separation between the stochastic components of the random function and its deterministic components.

7.2.1. *Monte Carlo (MC) method*

The theory of the MC method is presented in this section. [SPA 08] have given an exposition of this approach. This method provides successive solutions of a deterministic system that includes random variables. This method is used when the problem to be addressed is complex. It generates random samples for all uncertain parameters according to their probability distributions [GUR 15]. For each sample, a set of parameters is obtained and a deterministic calculation followed by a numerical or analytical model is carried out. The main advantage of this technique is its ease of implementation. Regardless of the size and complexity of a system, the results of such a method are accurate. It is nevertheless often expensive, since a reasonable accuracy requires a large number of samples.

In the present work, the MC method is considered as a reference method to validate the efficiency of GPC.

The functions of the MC method are as follows:

$$Y = M(X) \tag{7.1}$$

where X denotes the vector of uncertain input parameters modeled by a random vector $X = [X_1; X_2; ...; X_n]^T$, M is the model considered, and Y represents the vector of the estimated outputs, which will be a random vector.

7.2.2. Generalized polynomial chaos (GPC) method

This section describes the GPC method. An exposition of this method is given by [SEP 15]. The GPC provides, for any stochastic process, a linear combination of orthogonal polynomials of random variables. The main idea of the GPC method was developed by Wiener and extended by Xiu and Karniadakis [XIU 03], who used GPC to deal with general random fields such as multidimensional random fields and non-Gaussian random fields.

Consider a set of random variables (X_i) that can be written as a second-order random process. Consequently, ($X_{i,j}$) can be expressed in the following form:

$$X_{i,j}(x,\xi) = \sum_{j=0}^{\infty} Xb_{i,j}(x)\ \psi_j(\xi) \tag{7.2}$$

where ξ is the vector of random variables, $Xb_{i,j}$ is the deterministic component, and $\psi_j(\xi)$ are the orthogonal polynomial functions such as Legendre and Hermite.

Orthogonal polynomials verify the orthogonality relation:

$$<\psi_i,\psi_j> = \int \psi_i\ \psi_j\ W(\xi)\ d\xi = \begin{cases} 0 \text{ if } & i \neq j \\ <\psi_i,\psi_j> \text{ if } & i = j \end{cases} \tag{7.3}$$

where W (ξ) is the probabilistic density function and <,> is the internal product operator. In practice, the expansion of GPC is truncated to a finite order P in order to limit the number of terms of the sum to a finite number. This truncation is essential for the problem to remain feasible:

$$X_{i,j}(x,\xi) = \sum_{j=0}^{P} Xb_{i,j}(x)\ \psi_j(\xi) \tag{7.4}$$

with

$$P = \frac{(p+r)!}{p!r!} - 1 \qquad [7.5]$$

such that p is the order of polynomial chaos and r is the dimension of the random vector in the system. The calculation of the representation by the GPC method requires the determination of P+1 deterministic components called stochastic modes. Two approaches are needed to determine such modes: intrusive and non-intrusive. The first approach consists of substituting random functions, in the stochastic model, by their spectral representation. Then, a Galerkin projection of the resulting system on GPC is made [DAM 19b]. The second approach consists of projecting the stochastic solution directly from the GPC. A non-intrusive approach may be more attractive for a stochastic system since it requires few computations for deterministic solutions at a finite number of points in the probabilistic range of variation.

7.3. Deterministic design optimization

Deterministic design optimization (DDO) has become frequently used in structural design with the development of finite element (FE)-based software. This method consists of adding safety factors to the constrained function in view of integrating uncertainties into the design variables [YAI 17]. Thereby, the optimization problem consists of minimizing the objective function $f(x)$ subject to different constraints such as geometric, physical or functional constraints:

$$\begin{aligned} &\min_{x} : f(x) \\ &t.q : g_k(x,\xi) \leq 0 \quad k = 1,\ldots,K \end{aligned} \qquad [7.6]$$

where x is the vector of the deterministic design variables, ξ is the vector of the random variables, and $g_k(x, \xi)$ are the inequality constraints. In reality, designers cannot choose the appropriate safety factors to get as much safety as possible.

In order to overcome the problem of DDO, RBDO is essential to ensure a structural design that is both optimal and reliable.

7.4. Reliability-based design optimization (RBDO)

The RBDO procedure aims for the best trade-off between cost and security, taking into account the uncertainties associated with the system. The basic formulation of RBDO consists of minimizing the objective function under probabilistic constraints. In RBDO, two types of variables are taken into consideration: design variables x and random variables y. The typical RBDO formulation is written as [ELH 13]:

$$\min_{x} : f(x)$$
$$t.q: \begin{cases} \Pr\left[G_i(x,y) \leq 0\right] \leq P_{f_i}^T & i = 1,\ldots,r \\ h_k(x) \leq 0 & k = r+1,\ldots,R \end{cases} \quad [7.7]$$

where x is the vector of the design variables, y is the vector of the random variables, $f(x)$ is the objective function, $G_i(x,y)$ is the performance function, $h_k(x)$ are the deterministic constraints, Pr[.] is the probability operator, $P^T{}_{fi}$ is the probability of admissible failure, r is the number of execution functions, and R is the total number of constraints.

It should be noted that design variables x can be deterministic variables, system parameters (such as, for example, materials, loads, dimensions), or probabilistic distribution parameters (such as, for example, mean values and standard deviations of random variables y) [MAK 16]. Random variables y can be geometric dimensions or applied loads. The uncertainties of these variables can be modeled by statistical information. The deterministic constraints h_k are the upper and lower bounds of the design variables.

In the RBDO approach, the reliability index is iteratively computed. Consequently, solving an RBDO problem requires a lot of computational effort to solve such a design. To reduce the computation time, the hybrid method (HM) has been developed. It aims to simultaneously solve both the optimization and reliability evaluation. In doing so, the optimization problem becomes more complex than the deterministic problem. An effective method called optimal safety factor (OSF) has been developed. Through the use of this method, the computation time is reduced compared to that of the HM.

7.4.1. *Traditional method*

In the traditional approach, we use two spaces, namely the physical space and the normalized space [ELH 13]. Consequently, traditional RBDO is achieved by considering two interlinked problems:

– Optimization problem

Its aim is to minimize the objective function $f(x)$ subject to both the deterministic constraints $g_l(x)$ and the reliability constraint. It can be defined as follows:

$$\begin{aligned}&\min_{x}: \ f(x)\\&t.q:\begin{cases}g_l(x) \le 0 \quad l=0..L\\ \beta(x,u) \ge \beta_t\end{cases}\end{aligned} \qquad [7.8]$$

where $f(x)$ is the objective function, $g_l(x)$ are the associated constraints, $\beta(x, u)$ is the reliability index of the structure, and β_t is the reliability target.

– Reliability problem

The reliability index $\beta(x, u)$ is the minimal distance between the limit state function $H(x, u)$ and the origin. This index is determined by solving the minimization problem:

$$\begin{aligned}&\min_{u}: \ d_\beta(u)\\&t.q: \ H(x,u) \le 0\end{aligned} \qquad [7.9]$$

where $d_\beta(u)$ is the distance in normalized random space, given by $d_\beta(u) = \sqrt{\Sigma u_i^2}$ and $H(x,u)$ is the performance function (or limit state function) in normalized space.

It has been shown that the traditional approach requires a long computation time due to the large number of computations in both spaces and can lead to a low stability of the convergence. To overcome this problem, a new method called optimal safety factor (OSF) has been developed [KHR 08].

7.4.2. *Optimal safety factor (OSF) using GPC*

In the case of DDO, designers choose the safety factors based on engineering experience. Design safety is therefore not guaranteed, and the optimal point cannot lead to the required reliability level. Nevertheless, in the OSF method, the determination of safety factors is independent of the engineering experience. Therefore, the OSF approach is the best method to address this problem and to also reduce the computation time [ELH 13].

The OSF algorithm using the GPC method can be implemented in three steps. The proposed approach is described in the flowchart in Figure 7.1.

a) Determine the design point:

This step consists of solving the optimization problem by minimizing the objective function subject to the limit state function and the deterministic and probabilistic constraints calculated with the CPG. The resulting solution is designed by the design point, also called the most likely point of failure, and called the design point y^*_i.

b) Calculate the safety factors:

The design point y^*_i can be related to the optimal value x^*_i by equation [7.10]:

$$y_i^* = S_f^i \, x_i^*, \quad i=1,...,n \tag{7.10}$$

Moreover, in general, when considering the normal distribution of the random variable y_i, the transformation from the physical space (space x) to the normalized space (space u) is given by:

$$u_i = \frac{y_i - x_i}{\sigma_i}, \quad i=1,...,n \tag{7.11}$$

where σ_i is the standard deviation. In this way, we define the safety factors S^i_f, which can be expressed by:

$$S_f^i = 1 + \gamma_i \, u_i^* \quad i=1,..,n \tag{7.12}$$

where γ_i is the coefficient relating the standard deviation and the mean value of the random variable m_i. Therefore, γ_i can be expressed as follows:

$$\gamma_i = \frac{\sigma_i}{m_i} \tag{7.13}$$

The equation for the optimal value of the normalized vector u^*_i can be written as follows:

$$u_i^* = \pm \beta_t \sqrt{\frac{\left|\frac{\partial G}{\partial y_i}\right|}{\sum_{i=1}^{n}\left|\frac{\partial G}{\partial y_i}\right|}} \quad i = 1,\ldots,n \qquad [7.14]$$

The ± sign depends on the sign of the derivative:

$$\frac{\partial G}{\partial y_i} > 0 \rightarrow u_i^* > 0 \text{ and } \frac{\partial G}{\partial y_i} < 0 \rightarrow u_i^* < 0 \quad i = 1,..,n \qquad [7.15]$$

c) Calculate the optimal solution:

In this final step, we determine the optimal design by including the calculated safety factors in the values of the design variables.

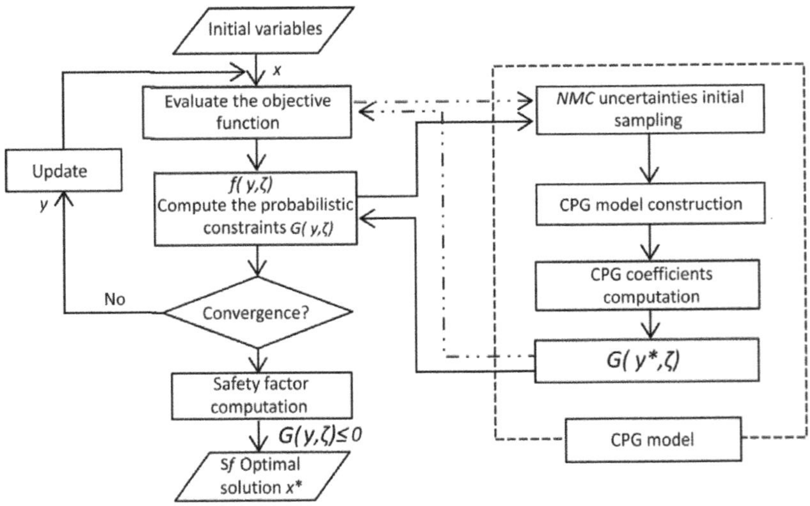

Figure 7.1. *Flowchart of the OSF function using the CPG meta-model*

The optimization problem is solved by the sequential quadratic programming (SQP) tool using MATLAB© software.

7.5. Numerical result

In this section, a numerical example of a 2D dental implant is studied. A probabilistic analysis is implemented to study the robustness of the system. Then, RBDO using GPC is performed in order to obtain the optimal structural design.

7.5.1. *2D dental implant*

Figure 7.2(a) shows a simplified geometric model of the dental implant under study. We consider two variables r and a which respectively represent the radius of curvature at the level of the pointed part of the implant and the radius of the implant. The material of the implant is a titanium alloy. The dimensions and properties can be found in Table 7.1. The numerical simulation of the dental implant is performed using the ANSYS© software. Concerning the discretization of the structure, 2,323 elements are adopted for the structure.

The boundary conditions adopted respectively are the F_x and F_y forces applied in the upper part of the dental implant, as shown in Figure 7.2(b). All nodes on the edge of the structure are constrained.

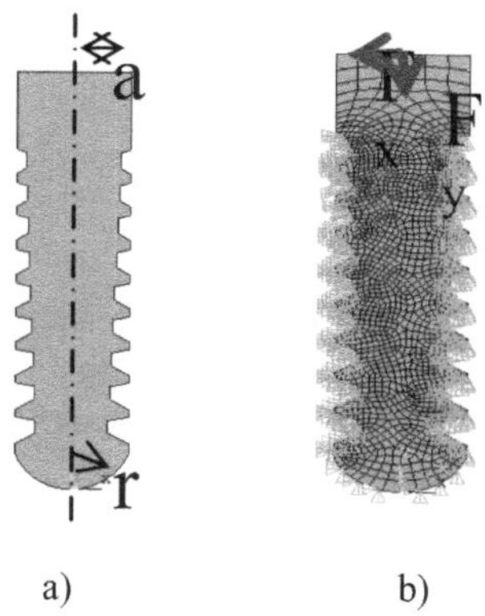

Figure 7.2. *a) Geometric model of the dental implant. b) Boundary conditions applied to the dental implant. For a color version of this figure, see www.iste.co.uk/elhami/multi-physics.zip*

Description	Parameters	Value
Curvature radius	r	2.1 mm
Implant radius	a	1.83 mm
Young's modulus	E	193,000 MPa
Poisson's ratio	v	0.31
Axial force	F*x*	30 N
Transversal force	F*y*	30 N

Table 7.1. *System dimensions and properties*

At the initial point, the initial values of the optimization variables are as follows: r = 2.1 mm and a = 1.83 mm, which lead to an initial value of the structural volume equal to $V_{initial} = 55.68\ mm^3$. The maximum value of the von Mises stress is equal to $\sigma_{max} = 177.18$ MPa.

For the optimization problem, the objective is to minimize the volume V of the dental implant subjected to the probabilistic constraint related to the admissible constraint condition. We consider that the design points are associated with the radius of curvature and the radius of the implant.

To perform the deterministic optimization, a safety factor $S_f = 1.25$ is proposed. The mathematical model of the DDO is given by:

$$\begin{aligned} &\min_{x} : V(x,\xi) \\ &t.q : \sigma_{\max}(x,\xi) \leq \frac{\sigma_w}{S_f} \end{aligned} \quad [7.16]$$

where σ_{max} is the maximum value of the von Mises stresses and σ_w is the admissible constraint equal to 185 MPa. To perform the reliability optimization by the OSF method, the first step consists of obtaining the design point. The formulation of the optimization problem is given as:

$$\begin{aligned} &\min_{x} : \mathrm{V}(x,\xi) \\ &t.q : \sigma_{\max}(x,\xi) \leq \sigma_w = 185 MPa \end{aligned} \quad [7.17]$$

The next step consists of calculating the safety factors and finally determining the optimal solution by dividing the value of the design point by the calculated safety factor.

7.5.1.1. *Probabilistic analysis*

In this section, the variability of uncertain parameters in the dental implant is introduced. The objective of this section is to improve the robustness and performance of the dental implant in the presence of uncertain parameters associated with the material properties.

These parameters are randomly chosen and have to follow a uniform distribution around their nominal 1% values. Table 7.4 presents an idea of the different uncertain parameters chosen for the example of the implant dental technique.

Parameters	E (MPa)	v
Value	U [191070,194930]	U [0.3069,0.3131]

Table 7.2. *Characteristics of uncertain parameters*

Dealing with the MC method consists of creating a grid of numerical values from the uncertain parameters and then calculating the quantity of interest of the system. The quantity of interest is analyzed for 1,000 simulations. Figure 7.3 shows the distribution of the input variables (E, ν) for a uniform distribution.

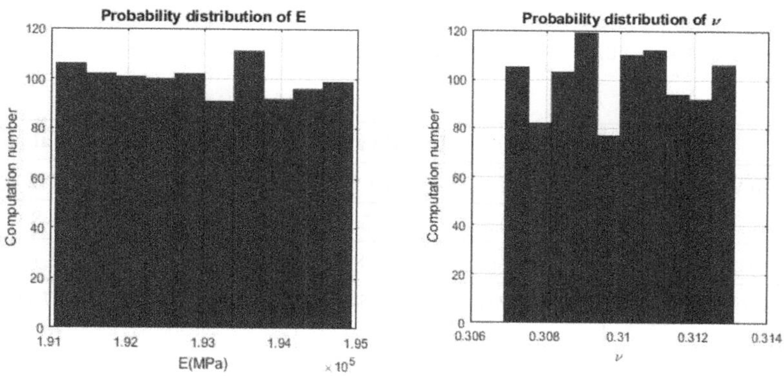

a) b)

Figure 7.3. *a) Dental implant geometric model. b) Boundary conditions applied to the dental implant. For a color version of this figure, see www.iste.co.uk/elhami/multi-physics.zip*

7.5.1.2. *RBDO using GPC*

To obtain the best structural design, we introduce the RBDO approach combined with FE simulations of the dental implant.

7.5.1.2.1. DDO approach using the GPC model.

Figure 7.5 shows the evolution of the objective function over the iterations.

The results of this technique are presented in Table 7.3. It is necessary to mention that the optimal solution satisfies the required constraint of the von Mises constraint, but the established reliability level is less than the reliability level $\beta_t = 3$.

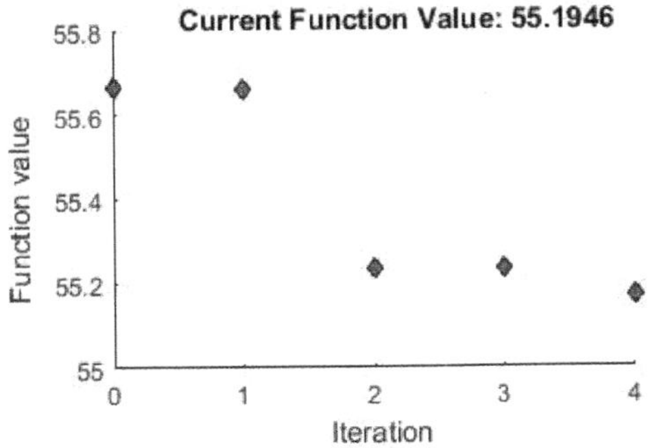

Figure 7.4. *Iterative curve of the volume of the dental implant by the DDO method. For a color version of this figure, see www.iste.co.uk/elhami/multi-physics.zip*

		Initial values	Design point	Optimal point
Design variables	r (mm)	2.1	2.13	2.12
	a (mm)	1.83	1.8	1.8
Objective function	V (mm^3)	55.68	55.17	55.19
Optimization constraint	σ_{max} (MPa)	179.54	175.60	177.18
Reliability index	β	-	-	1.18

Table 7.3. *DDO using GPC results*

This is the reason why the following method is used to ensure the robustness and reliability of our system.

7.5.1.2.2. DDO approach using the GPC model

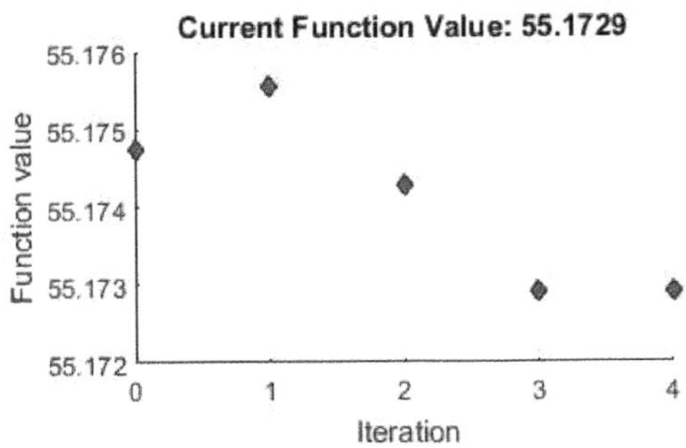

Figure 7.5. *Iterative curve of the volume of the dental implant by the OSF method. For a color version of this figure, see www.iste.co.uk/elhami/multi-physics.zip*

Figure 7.5 shows the evolution of the objective function with iterations, and Table 7.4 presents the results of this approach.

		Initial values	**Design point**	**OSF Sf**	**Optimal point**
Design variables	r (mm)	2.13	2.13	0.84	2.52
	a (mm)	1.8	1.8	1.04	1.72
Objective function	V (mm^3)	55.17	55.17		53.75
Optimization constraint	σ_{max} (MPa)	175.60	175.60		169.91
Reliability index	B	-	-		3

Table 7.4. *OSF using GPC results*

It should be noted that with the OSF approach, the computational cost is low compared to other RBDO methods such as the hybrid approach. Table 7.5 shows a comparison of the optimal point between the DDO method and the OSF method using the GPC meta-model.

	DDO	**OSF**
r (mm)	2.12	2.52
a (mm)	1.8	1.72
Volume (V)	55.19	53.75
Optimization constraint	177.18	169.91
Reliability index	1.18	3

Table 7.5. *Comparing the optimal point between DDO and OSF*

From Table 7.5, it can be observed that:

– In the deterministic optimization using the CPG model, the maximum tension stress is 121.39, which satisfies the constraint condition (less than $\sigma_{max} = S_f$) and the structural volume is 55.19 but the reliability level is low (β= 1.18), which does not guarantee the required reliability objective.

– In the OSF using the GPC model, the maximum von Mises stress is 169.91, which satisfies the stress condition (less than 185 MPa) and the structural volume is 53.75.

7.6. Conclusion

In this chapter, we use the GPC method for dental implant solutions when uncertainties exist in the material properties. In our study, we propose to take into account the uncertainty propagation by coupling the meta-model based on GPC with the OSF method.

The numerical example showed the effectiveness and efficiency of the OSF based on the CPG model for the RBDO of the dental implant with random variables.

Appendices

Appendix 1

Binomial Distribution

In many engineering applications, events consisting of repeating tests can be formulated as follows: occurrence or non-occurrence, success or failure, good or bad, etc. Only two solutions are possible, representing the behavior of a random discrete variable. Moreover, if the events satisfy the additional condition of a Bernoulli sequence, that is, if they are statistically independent and the probability of occurrence or non-occurrence remains constant, these events can be mathematically represented by a binomial distribution.

If the probability of occurrence of an event for each test is p and the probability of its non-occurrence is (1-p), then the probability of x occurrences out of a total of n tests can be determined by the mass function of a polynomial distribution:

$$P(X = x, n|\ p) = \binom{n}{x} p^x (1-p)^{n-x}\ x = 0,1,2,\ldots,n$$

where p is the distribution parameter, and $\binom{n}{x} = \frac{n!}{x!(n-x)!}$ is the binomial coefficient, indicating the number of ways in which x occurrences out of a total of n trials are possible.

Appendix 2

Geometric Distribution

The first occurrence of an event is very important in engineering. If the event occurs in a Bernoulli sequence and p is the probability of occurrence at each trial, then the probability that the event will occur at the i-th attempt, implying that it has not yet happened in the previous trials, is given by the geometric distribution:

$$P(T=t)=p(1-p)^{t-1} \quad t=1,2,\ldots \qquad \text{[A2.1]}$$

Return period

Changes in wind speed, rainfall, flooding or seismic hazards at a specific location are usually expressed in terms of a return period. Let us presuppose that an event appears for the first time in a Bernoulli sequence after a period of T_1 years.

This event occurs again T_2 years after the first occurrence, again T_3 years after the second occurrence, etc. The recurrence time, namely the time between two consecutive occurrences of the same event, must follow the characteristic probability of the first occurrence, that is a geometric distribution whose mass probability is given by equation [A2.2].

The average return time, also known as the return period, can thus be calculated as:

$$\begin{aligned} T = E(T) &= \sum_{t=1}^{\infty} t p_T(t) = \sum_{t=1}^{\infty} t p(1-p)^{t-1} \\ &= p[1+2(1-p)+3(1-p)^2+4(1-p)^3+\cdots] \end{aligned} \qquad \text{[A2.2]}$$

However, the bracketed terms in equation [A2.3] represent an infinite series and can be described as $1/p^2$. Therefore, equation [A2.3] can be written in the following simplified manner as:

$$T = p \times \frac{1}{p^2} = \frac{1}{p} \qquad [A2.3]$$

Appendix 3

Poisson Distribution

The Poisson distribution is typically used to assess risks in the industry. It takes into account time or space problems. If we prefer to solve this kind of problem with the Bernoulli distribution, the interval of time or space should be broken down into very small parts, because it only accounts for the occurrence or absence of an event.

Suppose that the average occurrence of a tornado at a given place is v times a year; during a period of t years, a tornado will occur on average vxt times. If the time period is divided by n intervals, the probability of a tornado occurring will be (vxt)/n. If we take x occurrences during a period of time t in a Bernoulli sequence when n tends to infinity, we then meet the Poisson distribution, which can be expressed as follows:

$$\begin{aligned}
P(x \text{ occurrences in } t) &= \lim_{n\to\infty}\binom{n}{x}\left(\frac{vt}{n}\right)^{x}\left(1-\frac{vt}{n}\right)^{n-x} \\
&= \lim_{n\to\infty}\left[\frac{n}{n}\,\frac{(n-1)}{n}\cdots\frac{(n-x+1)}{n}\cdots\frac{(vt)^{x}}{x!}\left(1-\frac{vt}{n}\right)^{n}\left(1-\frac{vt}{n}\right)^{-x}\right] \\
&= \lim_{n\to\infty}\left[\frac{(vt)^{x}}{x!}\left(1-\frac{vt}{n}\right)^{n}\right]
\end{aligned}$$

The limit of this equation is taken and knowing that:

$$\lim_{n\to\infty}\left(1-\frac{vt}{n}\right)^{n} = 1 - vt + \frac{(vt)^2}{2!} - \frac{(vt)^3}{3!} + \cdots = e^{-vt}$$

It can therefore be shown that:

$$P(x \text{ occurrences in a time } t) = \frac{(v \times t)^x}{x!} e^{-v \times t}$$

Appendix 4

Exponential Distribution

If an event follows a Poisson logic, then the time T before the first occurrence of that event can be represented by an exponential distribution, as follows:

$$P(T > t) = \frac{e^{\nu \times t} (\nu \times t)^0}{0!} = e^{-\nu \times t}$$

thereby the distribution function is written as:

$$F_T(t) = P(T \leq t) = 1 - e^{-\nu \times t}$$

and the probability density function of the exponential distribution is written as:

$$f_t(t) = \frac{dF_T(t)}{dt} = \nu \times e^{-\nu \times t}, \quad t \geq 0.$$

It can be shown that the mean value of T is equal to 1/v. That is, the mean time of the first occurrence, or the time of recurrence, or simply the return period for a Poisson model is 1/v. Interestingly, it should be noted that when an event follows a Bernoulli model, the return period is 1/p. It can be shown that for an event with a low occurrence v, the return periods for a Bernoulli sequence and for a Poisson model are approximately identical.

Appendix 5

Normal Distribution

A real random variable X follows a normal, or Gaussian, distribution of mean μ and standard deviation σ if its probability density f is given by:

$$\begin{cases} f(x) = \dfrac{1}{\sigma\sqrt{2\pi}} e^{-\frac{1}{2}\left(\frac{x-\mu}{\sigma}\right)^2} \\ \text{with } x \in \mathbb{R},\ \mu \in \mathbb{R},\ \sigma \in \mathbb{R}^{+*} \end{cases}$$

This function is determined by knowing only the parameters μ and σ. We write: $X \sim \mathcal{N}(\mu,\sigma) : X$ follows the distribution $\mathcal{N}(\mu,\sigma)$.

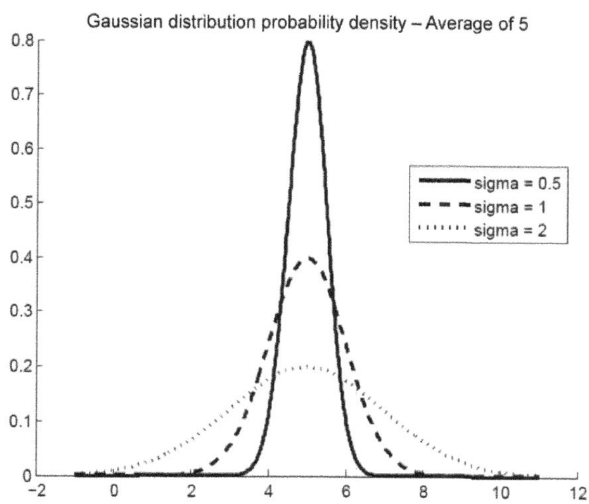

Figure A5.1. *Influence of the standard deviation with a constant mean*

The graphical representation of f(x) gives rise to the famous "bell curve".

Regardless of the parameter values, the Gaussian distribution exhibits a maximum amplitude called the statistical mode. The mean μ corresponds to the abscissa of this maximum amplitude. The intervals classically considered are: [μ-σ,μ+σ], [μ-2σ,μ+2σ], [μ-3σ,μ+3σ] and [μ-4σ,μ+4σ], which respectively contain 68.3, 95.5, 99.7 and 99.99% of the statistical population considered. The distribution function F is defined by:

$$F(x)=\int_{-\infty}^{x}(t)dt=\int_{-\infty}^{x}\frac{1}{\sigma\sqrt{2\pi}}e^{-\frac{1}{2}\left(\frac{t-\mu}{\sigma}\right)^2}dt$$

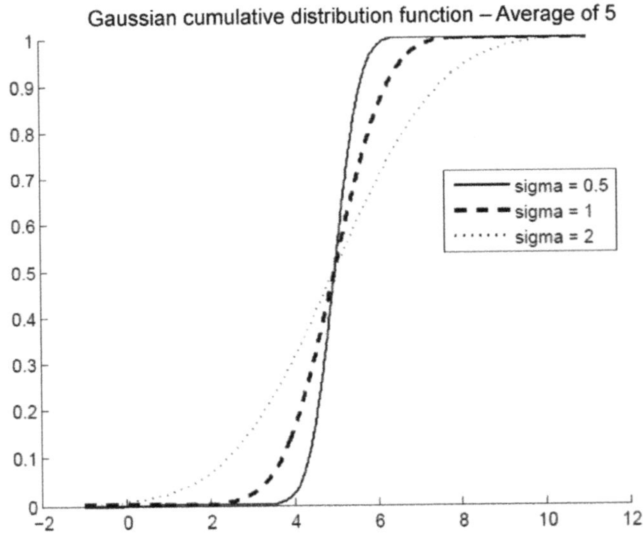

Figure A5.2. *Influence of a standard deviation with a constant mean*

Reduced centered normal distribution

A Gaussian random variable X_0 with a zero mean and a unit standard deviation follows the reduced centered normal distribution (μ = 0 and σ = 1): $X_0 \sim \mathcal{N}(0,1)$. Its associated probability density directly follows from:

$$f(x_0)=\frac{1}{\sqrt{2\pi}}e^{-\frac{x_0^2}{2}} \quad x_0 \in \mathbb{R}$$

Conventionally, the distribution function of the reduced centered normal distribution is denoted by ϕ:

$$\Phi(x_0) = \int_{-\infty}^{x_0} f(t_0)\,dt_0 = \int_{-\infty}^{x_0} \frac{1}{\sqrt{2\pi}} e^{-\frac{1}{2}t_0^2}\,dt_0$$

If $X \sim \mathcal{N}(\mu,\sigma)$ then $X_0 = \dfrac{X-\mu}{\sigma} \sim \mathcal{N}(0,1)$ and therefore:

$$\Phi\left(\frac{x-\mu}{\sigma}\right) = \int_{-\infty}^{\frac{x-\mu}{\sigma}} \frac{1}{\sqrt{2\pi}} e^{-\frac{1}{2}\left(\frac{t-\mu}{\sigma}\right)^2} \frac{dt}{\sigma} = \int_{-\infty}^{x_0} \frac{1}{\sigma\sqrt{2\pi}} e^{-\frac{t_0^2}{2}}\,dt = F(x)$$

It then follows that

$$E(X_0) = E\left(\frac{X-\mu}{\sigma}\right) = \frac{1}{\sigma}E(X) - \frac{\mu}{\sigma} = \frac{\mu-\mu}{\sigma} = 0$$

$$V(X_0) = \frac{E(X^2)-\mu^2}{\sigma^2}$$

$$\sqrt{V(X_0)} = \frac{\sqrt{E(X^2)-\mu^2}}{|\sigma|} = \frac{\sqrt{V(X)}}{\sigma} = \frac{\sigma}{\sigma} = 1$$

where E is the mean, and V is the variance of the variables considered.

The odd-order moments of the reduced centered normal distribution are all zero. This is due to the parity of the probability density.

The moments of order n and 2k+1 are written as:

$$m_n = \int_{-\infty}^{+\infty} t^n \frac{1}{\sqrt{2\pi}} e^{-\frac{t^2}{2}}\,dt$$

$$m_{2k+1} = \int_{-\infty}^{+\infty} t^{2k+1} \frac{1}{\sqrt{2\pi}} e^{-\frac{t^2}{2}}\,dt = 0$$

and even-order moments:

$$m_{2k} = \int_{-\infty}^{+\infty} t^{2k} \frac{1}{\sqrt{2\pi}} e^{-\frac{t^2}{2}} dt$$

It is easy to show that:

$$m_{2k+2} = (2k+1)\, m_{2k}$$

Now, $m_0 = 1$, hence: $m_{2k} = \dfrac{(2k)!}{2^k k!}$

Appendix 6

Log-Normal Distribution

A random variable Y follows the log-normal distribution of parameters μ and σ if its natural logarithm follows the normal distribution:

$$Y \sim \ln-\mathcal{N}(\mu,\sigma)$$
$$X = \ln(Y) \sim \mathcal{N}(\mu,\sigma)$$

The associated probability density is:

$$f(y) = \frac{1}{\sigma y\sqrt{2\pi}} e^{-\frac{1}{2}\left(\frac{\ln(y)-\mu}{\sigma}\right)^2} \quad \text{with } y \in \mathbb{R}^{+*}, \mu \in \mathbb{R}, \sigma \in \mathbb{R}^{+*}$$

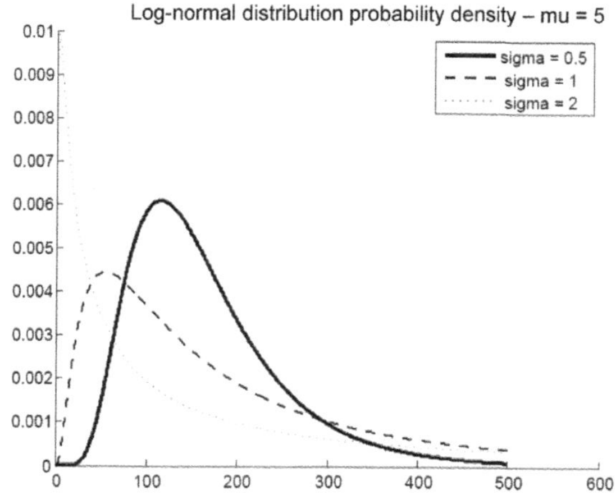

Figure A6.1. *Influence of a standard deviation with a constant mean*

The distribution function F is given by:

$$F(y) = \Phi\left(\frac{\ln(y) - \mu}{\sigma}\right)$$

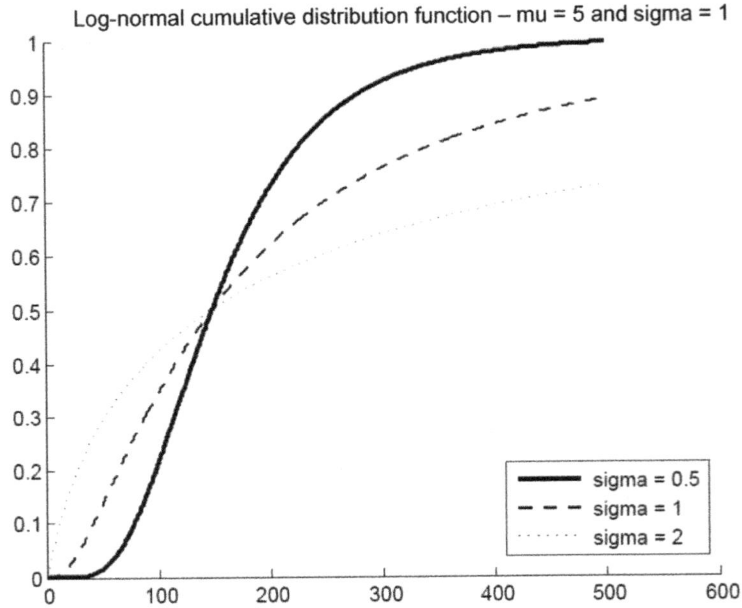

Figure A6.2. *Influence of a standard deviation with a constant mean*

The k-order moments of the log-normal distribution are:

$$m_k = \int_0^{+\infty} y^k f(y)\,dy = \int_0^{+\infty} y^k \frac{1}{\sigma y \sqrt{2\pi}} e^{-\frac{1}{2}\left(\frac{\ln(y)-\mu}{\sigma}\right)^2} dy$$

After calculation, we find that: $m_k = e^{k\mu + \frac{k^2\sigma^2}{2}}$

The mean is given by:

$$E(Y) = m_1 = e^{\mu + \frac{\sigma^2}{2}}$$

The variance is given by:

$$V(Y) = e^{2\mu+\sigma^2}\left(e^{\sigma^2} - 1\right)$$

from which the standard deviation is written as:

$$\sigma_Y = \sqrt{V(Y)} = e^{\mu+\frac{\sigma^2}{2}}\sqrt{e^{\sigma^2} - 1}$$

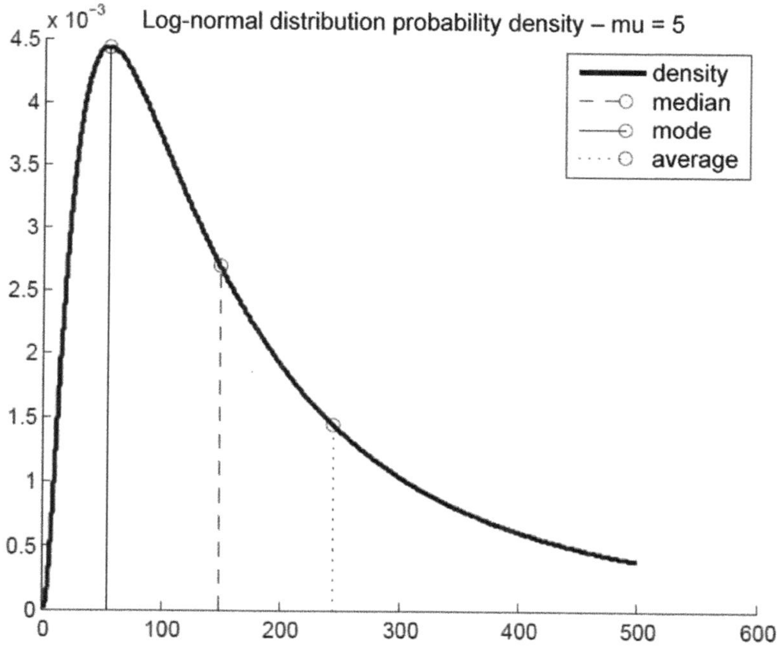

Figure A6.3. *Position of the median, mode and mean – to be compared with μ*

A property of the log-normal distribution is that, unlike the normal distribution, the mean, the median and the mode are not mistaken for one another.

Appendix 7

Weibull Distribution

The Weibull distribution includes a whole family of continuous probability distribution generally applied to the lifetimes of many types of equipment (systems, spare parts, relays, fatigue, wear, etc.). In reliability-based control, companies tend to use it, and more precisely, when the failure rate evolves as a power of time. When this rate is constant, the exponential distribution, a particular form of the Weibull distribution, is used. The Weibull distribution is based on three parameters, which are as follows:

– α: time scale parameter;

– β: shape parameter;

– γ: tracking or localization parameter.

These parameters, whose influence will be studied later, will allow us to modify the different periods (see Figure A7.1).

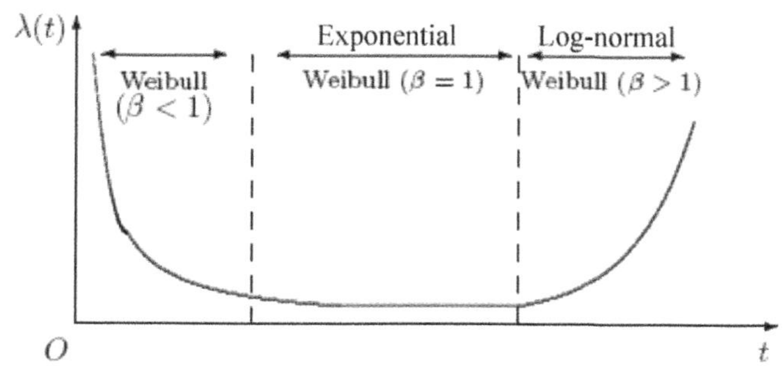

Figure A7.1. *The Weibull distribution*

The different indicators are as follows:

a) Damage rate: Weibull has proposed the following model for the damage rate:

$$a(t) = \frac{d}{dt}\frac{(t-\gamma)^{\beta}}{\alpha} = \frac{\beta}{\alpha}(t-\gamma)^{\beta-1}$$

b) Reliability: R(t) is the probability of the proper functioning at time t.

By definition, we have:

$$a(t)dt = -\frac{dR(t)}{R(t)}.$$

which leads to:

$$\begin{cases} R(t) & = & \exp(-\int_0^t a(x)dx) \quad t \in [0,+\infty] \\ R(0) & = & 1 \end{cases}$$

However:

$$\int_0^t a(x)dx = \int_0^t \frac{d}{dx}\left[\frac{(x-\gamma)^{\beta}}{\alpha}\right]dx = \frac{(t-\gamma)^{\beta}}{\alpha}.$$

Finally, we obtain:

$$R(t) = \exp\left(-\frac{(t-\gamma)^{\beta}}{\alpha}\right) = \exp\left(-\left(\frac{t-\gamma}{\eta}\right)^{\beta}\right) \quad (\eta^{\beta} = \alpha)$$

We write: $\eta^{\beta} = \alpha$.

c) Distribution function: from the reliability R(t), the distribution function F(t) is directly obtained:

$$F(t) = 1 - R(t) = 1 - \exp\left(-\left(\frac{t-\gamma}{\eta}\right)^{\beta}\right)$$

d) Probability density: we have by definition:

$$f(t) = \frac{d}{dt}F(t) = \frac{d}{dt}\left(1 - e^{\left(-\left(\frac{t-\gamma}{\eta}\right)^{\beta}\right)}\right) = \frac{\beta}{\eta}.\left(\frac{t-\gamma}{\eta}\right)^{\beta-1}.e^{\left(-\left(\frac{t-\gamma}{\eta}\right)^{\beta}\right)}$$

e) Instantaneous failure rate: by definition, we have:

$$\lambda(t) = \frac{f(t)}{R(t)} = \frac{\frac{\beta}{\eta}.\left(\frac{t-\gamma}{\eta}\right)^{\beta-1}.e^{\left(-\left(\frac{t-\gamma}{\eta}\right)^{\beta}\right)}}{e^{-\left(\frac{t-\gamma}{\eta}\right)^{\beta}}}$$

$\lambda(t)$ is a reliability estimator.

f) Lifetime: the lifetime is determined from R(t). We have:

$$R(t) = e^{-\left(\frac{t-\gamma}{\eta}\right)^{\beta}}$$

from which:

$$\ln(R(t)) = -\left(\frac{t-\gamma}{\eta}\right)^{\beta} \Rightarrow \ln\left(\frac{1}{R(t)}\right) = \left(\frac{t-\gamma}{\eta}\right)^{\beta} \Rightarrow \ln\left(\frac{1}{R(t)}\right)^{\frac{1}{\beta}} = \frac{t-\gamma}{\eta}$$

Finally, we obtain:

$$t = \eta.\ln\left(\frac{1}{R(t)}\right)^{\frac{1}{\beta}} + \gamma.$$

g) Mean: the expression of the mean is:

$$\bar{x} = \alpha^{\frac{1}{\beta}}.\frac{1}{\beta}.\Gamma\left(\frac{1}{\beta}\right) = \eta.\frac{1}{\beta}\Gamma\left(\frac{1}{\beta}\right)$$

Recall that the function Γ is defined by: $\Gamma(x) = \int_0^{+\infty} t^{x-1} e^{-t} dt$. The function Γ is described in the statistical tables.

h) Variance: the expression of the variance is given from the standard deviation σ by:

$$\sigma^2 = \eta^2 \left(\Gamma\left(\frac{2}{\beta}+1\right) - \Gamma\left(\frac{1}{\beta}-1\right)^2 \right).$$

Appendix 8

Pareto Distribution

The Pareto variable, with parameter α and t_0 ($t_0 \geq 0$), is the random variable t whose distribution function is given by:

$$\begin{cases} F(t) & = & 0 & \text{if } t < t_0 \\ F(t) & = & 1-\left(\dfrac{t_0}{t}\right)^{\alpha} & \text{if } t \geq t_0 \end{cases}$$

with

$$\begin{aligned} P\{t \text{ lifespan} < t\} &= F(t) \\ P\{t \text{ lifespan} \geq t\} &= R(t) = 1 - F(t) \end{aligned}$$

The density of this variable is given by:

$$f(t) = \frac{dF(t)}{dt} = \frac{d(-t_0^{\alpha} \times t^{-\alpha})}{dt}$$

then

$$f(t) = \frac{\alpha}{t_0} \cdot \left(\frac{t_0}{t}\right)^{\alpha+1}$$

There are two characteristics in the Pareto principle: quantiles and moments. We then have:

1) Quantiles:

$$F(t_p) = 1 - \left(\frac{t_0}{t_p}\right)^{\alpha} = p$$

Therefore:

$$t_p = \frac{t_0}{(1-p)^{1/\alpha}}$$

The median is given by: $M_e = t_{0,5} = t_0.2^{1/\alpha}$.

2) Moment: the non-centered moment of order r is m_r; it is calculated as follows:

$$m_r = \int_{t_0}^{+\infty} f(t)t^r dt = \frac{\alpha t_0^r}{\alpha - r} \quad \text{if } \alpha < r$$

The average $\overline{t}$ exists only if $\alpha > 1$:

$$\overline{t} = m_1 = \frac{\alpha}{\alpha - 1}.t_0$$

The variance V exists only if $\alpha > 2$:

$$V = m_2 - m_1^2 = \frac{\alpha.t_0^2}{\alpha - 2} - \left(\frac{\alpha}{\alpha - 1}t_0\right)^2$$

Therefore:

$$V = \frac{\alpha.t_0^2}{(\alpha - 2)(\alpha - 1)^2}$$

The standard deviation $\sigma(t)$ is:

$$\sigma(t) = \frac{t_0}{\alpha - 1}.\sqrt{\frac{\alpha}{\alpha - 2}}$$

Appendix 9

Extreme Value Distribution

In many engineering applications, extreme values of random variables have an important significance. For example, the highest or lowest value can indicate a particular pattern. Take the example of wind speed, which is continuously recorded at airports or weather stations. Obviously, the very large amount of data cannot be used as it is in mathematical models. The maximal wind speed per hour, day, month, year, or any other period can be used for this purpose. Information about the maximal wind speed over a year is usually taken into account in applications. Therefore, for each batch of data covering a year, the maximal wind speed is recorded. If the data is taken for many years, the wind speed model can be statistically established to be sure that it will not exceed the lifespan for which the structure was designed for.

If the wind speed model has a return period of 50 years, then the probability that the wind speed will exceed the nominal speed over a year is 1/50 = 0.02. Earthquakes, floods and other climatic events can thus be determined. In all of these cases, it is important to consider the maximal value of a random variable during a certain period of time. In some cases, the minimal value is also important to analyze. For example, when a large number of devices are manufactured, such as computers or cars, their minimum service life is an important aspect that consumers have to consider. Therefore, accelerated lifetime tests can be used to construct a probability of lifetime distribution. The minimal lifetime can be set such that it is not less than a value considered acceptable.

When building an extreme value distribution, it is necessary to establish an underlying random variable with a particular distribution. If different sets of samples are obtained (through physical or numerical experiments), the extreme values can be selected for each set of samples, and as such, construct a particular distribution for the extreme values.

Extreme value distribution concept

Let X be a random variable with an unknown distribution function. If there are n samples for the random variable X, it might be interesting to look into the boundary values for the samples, such as the minimal value Y_1 or the maximal value Y. Y_1 and Y_n can be defined as follows:

$$Y_n = \max(X_1, X_2, \ldots, X_n) \quad \text{and} \quad Y_1 = \min(X_1, X_2, \ldots, X_n) \tag{A9.1}$$

If different sets of samples of the same size n are obtained for X, each set will have different maximal and minimal values. Using all of these sets, distribution functions for minimal and maximal values can be constructed. The cumulative distribution function (CDF) of the largest value Y_n is:

$$F_{Y_n}(y) = P(Y_n \leq y) = P(X_1 \leq y, X_2 \leq y, \ldots, X_n \leq y) \tag{A9.2}$$

for uniformly distributed and statistically equivalent variables X_i, the previous equation becomes:

$$F_{Y_n}(y) = [F_X(y)]^n. \tag{A9.3}$$

Similarly, the CDF of the smallest value Y_1 is:

$$P(Y_1 > y) = P(X_1 > y, X_2 > y, \ldots, X_n > y) = 1 - F_{Y_1}(y) \tag{A9.4}$$

For uniformly distributed and statistically independent X_i, equation [A9.3] becomes:

$$F_{Y_1}(y) = 1 - [1 - F_X(y)]^n \tag{A9.5}$$

The main idea of this formula is that if the largest value Y_n is less than y, then all of the values of the sample (X_1, X_2, ...) must also be less than y, and the same applies for the smallest value Y_1.

Appendix 10

Asymptotic Distributions

The sample size grows a lot and tends to infinity, and the distribution of the largest or smallest values can asymptotically tend toward a mathematical distribution function, in some cases, if the samples are distributed in an identical and statistical independent manner. Some of these asymptotic distributions have a wide range of applications.

Gumbel (1958) listed three types of extreme value distributions known as asymptotic distributions, categorized as type I, type II, and type III. Type I extreme value distributions are widely used in mechanical reliability applications. The distribution of maxima in sample sets established from a population following a normal distribution asymptotically converges to this distribution. This type of distribution is used to model climatic phenomena such as wind or variations in water levels.

Type II distributions are also used to create models of environmental phenomena, such as earthquakes, and can be inferred with sample sets from a log-normal distribution.

Type III distributions, also known as the Weibull distribution in the case of smaller values, can be obtained through the convergence of most distributions that have a lower bound. It is often used to describe the mechanical strength of a material or the time an electronic or mechanical device properly operates.

Extreme value distributions can be addressed in the same way as other distributions, either defined in terms of their probability density function or distribution function and the associated parameters. In most cases, the parameters can be estimated from the information about the mean, variance or coefficient of variation of the random variable. Once an extreme value distribution is exclusively described, the probabilistic information can be extracted using the methods presented above.

Type I distribution

The type-I asymptotic CDF of the distribution of the largest value, also called the Gumbel distribution, can be expressed in the following manner:

$$F_{Y_n}(y_n) = \exp\left[-e^{\alpha_n(y_n - u_n)}\right]$$

where u_n is the largest characteristic value of the initial variable X, and α_n is the inverse measure of the dispersion of the largest value of X.

It can be shown that the probability distribution function is:

$$f_Y(y) = \alpha e^{\alpha_n(y_n - u_n)} \exp\left[-e^{\alpha_n(y_n - u_n)}\right], \quad -\infty < y < +\infty$$

U_n and α_n are related to the mean and standard deviation of the extreme variable Y_n:

$$\alpha_n = \frac{1}{\sqrt{6}}\left(\frac{\pi}{\sigma_{Y_n}}\right) \quad \text{and} \quad u_n = \mu_{Y_n} - \frac{0.5772}{\alpha_n}$$

The type-I extreme value distribution of the largest value is commonly used for modeling natural phenomena. It is a very important variable in the design of flood control, drinking water consumption and irrigation systems.

For the smallest value of a variable X, the corresponding form of the asymptotic type I for the CDF is:

$$F_{Y_1}(y_1) = 1 - \exp\left[-e^{\alpha_1(y_1 - u_1)}\right].$$

and the probability density function is:

$$f_{Y_1} = \alpha_1 e^{\alpha_1(y_1 - u_1)} \exp\left[-e^{\alpha_1(y_1 - u_1)}\right], \quad -\infty < y_1 < +\infty$$

Type II distribution

The distribution function for a type-II distribution of a maximum value, also called the Fréchet distribution, can be expressed as follows:

$$F_{Y_n}(Y_n) = \exp\left[-\left(\frac{v_n}{y_n}\right)^k\right]$$

The corresponding probability density function is:

$$f_{Y_n}(y_n) = \frac{k}{v_n}\left(\frac{v_n}{y_n}\right)^{k+1} \exp\left[-\left(\frac{v_n}{y_n}\right)^k\right], y_n \leq 0, k > 2$$

where v_n and k are the parameters of the distribution; v_n is the characteristic maximal value of the underlying variable X, and k, the shape parameter, is a dispersion measure.

The asymptotic form for a type II distribution is obtained when n goes to infinity from an initial distribution that has a polynomial tangency in the direction of the extreme value. We should observe the difference between this peculiarity and type I, which converges from an exponential tangency. The type-II distribution requires a polynomial direction, and consequently, a log-normal distribution converges to an asymptotic type-II form for maximal values.

Type III distribution

The type III in asymptotic distributions represents a distribution to know the upper or lower bounds of distributions that have finite limits. The CDF for the asymptotic type III can be described as follows:

$$F_{Y_n}(y_n) = \exp\left[-\left(\frac{\omega - y_n}{\omega - w_n}\right)^k\right]$$

and the probability density function as:

$$f_{Y_n}(y_n) \frac{k}{\omega - w_n}\left(\frac{\omega - y_n}{\omega - w_n}\right)^{k-1} \exp\left[-\left(\frac{\omega - y_n}{\omega - w_n}\right)^k\right], y_n \leq \omega$$

where ω is the upper bound of the initial distribution, that is, $F_X(w_n)=1-1/n$; w_n and k are the distribution parameters; w_n is the largest characteristic value of X and is defined by: $F_X(w_n)=1-1/n$, and k is the shape parameter. The mean and variance of Y_n are related to the parameters w_n and k as follows:

$$\mu_{Y_n} = \omega - (\omega - w_n)\Gamma\left(1+\frac{1}{k}\right)$$

and

$$\sigma_{Y_n}^2 = Var(\omega - Y_n) = (\omega - w_n)^2\left[\Gamma\left(1+\frac{2}{k}\right) - \Gamma^2\left(1+\frac{1}{k}\right)\right]$$

Based on the two previous equations, we show that:

$$1+\left(\frac{\sigma_{Y_n}}{\omega - \mu_{Y_n}}\right)^2 = \frac{\Gamma\left(1+\frac{2}{k}\right)}{\Gamma^2\left(1+\frac{1}{k}\right)}$$

The type-III CDF for the smallest value is defined as follows:

$$F_{Y_1}(y_1) = \exp\left[-\left(\frac{y_1-\epsilon}{w_1-\epsilon}\right)^k\right]$$

and the probability density function is written as:

$$f_{Y_1}(y_1)\frac{k}{w_1-\epsilon}\left(\frac{y_1-\epsilon}{w_1-\epsilon}\right)^{k-1}\exp\left[-\left(\frac{y_1-\epsilon}{w_1-\epsilon}\right)^k\right], y_1 \geq \epsilon$$

where ω is the upper bound of the initial distribution, that is, $F_X(w_n)=1-1/n$; w_n and k are the distribution parameters; w_n is the largest characteristic value of X and is defined by: $FX(w_1)=1-1/n$ and k is the shape parameter. The mean and variance of Y_1 are related to the parameters w_1 and k as follows:

$$\mu_{Y_1} = \epsilon + (w_n - \epsilon)\Gamma\left(1+\frac{1}{k}\right)$$

and

$$\sigma_{Y_1}^2 = Var(Y_1 - \epsilon) = (w_1 - \epsilon)^2 \left[\Gamma\left(1+\frac{2}{k}\right) - \Gamma^2\left(1+\frac{1}{k}\right) \right]$$

Based on the two previous equations, we show that:

$$1+\left(\frac{\sigma_{Y_1}}{\mu_{Y_1} - \epsilon}\right)^2 = \frac{\Gamma\left(1+\frac{2}{k}\right)}{\Gamma^2\left(1+\frac{1}{k}\right)}$$

Appendix 11

Introduction to Optimization

A11.1. Introduction

The term optimization is often associated with minimizing or maximizing a function. The only problem that has to be taken into consideration is that which involves minimizing one function $f(x)$ only. This is because the problem of maximizing the function $f(x)$ is simply a problem of minimizing the function $-f(x)$. In engineering problems, optimization is often linked to the design problem. The function $f(x)$ is known as the cost function or the objective function: it is the quantity that we want to be as small as possible such as, for example, cost and weight. The components of the vector x, known as the design variable, are quantities that ought to be fit. Physical dimensions (lengths, areas, angles, etc.) are common examples of design variables.

The purpose of this appendix is to introduce this field of optimization, whose literature is very rich, by defining this problem. The most widely used techniques to find the minimum of the function will be introduced. Minimization algorithms are iterative procedures that need an initialization vector x^0.

If the function $f(x)$ has several local minima, the choice of x^0 directs the algorithm toward the local minimum to be calculated. There is no guarantee that the overall minimum will be found. One idea that should be explored is to calculate different local minima and to choose from these minima the one that gives the best value of the function $f(x)$.

A11.2. Optimization without constraints

A11.2.1. *Quadratic functions*

DEFINITION.– A quadratic function is any function such as:

$$f(x) = \frac{1}{2} x^T Ax - b^T x \qquad \text{[A11.1]}$$

where $A \in \mathcal{M}_{nn}(\mathbb{R})$ is a symmetric matrix and $b \in \mathbb{R}^n$.

THEOREM.– A necessary and sufficient condition for the quadratic function defined by equation [A1.1] to be strictly convex is that A be definite positive.

REMARK A11.1.– A matrix A is said to be positive if and only if for all x in $\mathbb{R}^n_*$, one has $x^t Ax > 0$.

REMARK A11.2.– Characterization of convexity.

The function $f : \mathbb{R}^n \to \mathbb{R}$ is convex (respectively strictly convex) if and only if the Hessian matrix $\nabla^2 f(x)$ is semidefinite positive (respectively definite positive) for any $x \in \mathbb{R}^n$.

A11.2.2. *Existence and uniqueness results*

We have the following results:

THEOREM.– Existence in $\mathbb{R}^n$

Let f: $\mathbb{R}^n \to \mathbb{R}$ be a continuous and coercive function*. Then, there exists at least one element $\rho > 0$ such that

$$f(\overline{x}) \leq f(x) \quad \forall x \in \mathbb{R}^n$$

REMARK A11.3.– In practice, the properties of (strict) convexity are rarely verified. We can then narrow down a “local” aspect. We will speak of local minimum for $\overline{x}$ if there exists $\rho > 0$ such that:

$$\| x - \overline{x} \| \leq \rho \Rightarrow f(\overline{x}) \leq f(x)$$

A11.2.3. *Optimality conditions*

The following considers only the case without constraint. Consider a given continuously differentiable function $f : \mathbb{R}^n \to \mathbb{R}$. The optimization problem outlined consists of finding $\tilde{x} \in \mathbb{R}^n$ such that:

$$f(\tilde{x}) \le f(x) \quad \forall x \in \mathbb{R}^n \tag{A11.2}$$

THEOREM.– A necessary condition for $\tilde{x}$ to be the solution to the optimization problem equation [A11.2] is $\nabla f(\tilde{x}) = 0$.

REMARKS A11.4.–

– The previous theorem gives a necessary condition, which is generally not sufficient.

– The necessary condition remains valid for a local minimum.

THEOREM.– Let f be a convex function. Let $\overline{x} \in \mathbb{R}^n$ be such that $\nabla f(\overline{x}) = 0$. Then, $\overline{x}$ is the solution of the global optimization problem equation [A11.2].

EXAMPLE.– *Quadratic case*

Let $A \in \mathcal{M}_{nn}(\mathbb{R})$ be a symmetric matrix. Let $b \in \mathbb{R}^n$. Consider the quadratic function f defined by $f(x) = \frac{1}{2} x^T Ax - b^T x$. The necessary condition of optimality is written as $\nabla f(\overline{x}) = 0 \Leftrightarrow (A\overline{x} - b)^T = 0$.

Therefore, if A is positive definite, the unique minimum is given by equation $A\overline{x} = b$. It is therefore possible to reduce the problem of minimizing a quadratic function to the solution of a linear system.

A11.3. Solving methods

A11.3.1. *Golden section method*

The golden section method is used to calculate the minimum of a function $f(x)$ when n is equal to 1 without calculating the derivative of the function $f(x)$. It is assumed that the function f is continuous over the interval $[a, b]$ and that f

admits only one minimum in $[a, b]$. The idea of this method is the same as that of the bisection method.

We start by calculating the middle $c = \frac{a+b}{2}$ and evaluate x_1 and x_2 defined by: $x_1 = c - \delta/2$ and $x_2 = c + \delta/2$ for a small δ such that $f(x_1) \neq f(x_2)$. If $f(x_1) < f(x_2)$, then the search for the minimum carries on in the interval $[a, x_1]$, otherwise, the search switches to the interval $[x_2, b]$ (see Figure A11.1).

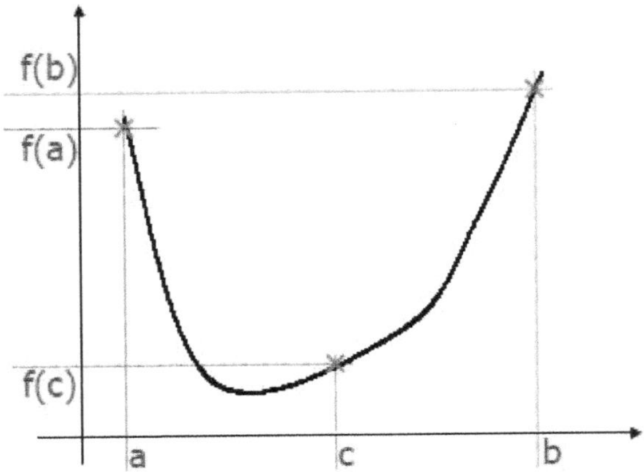

Figure A11.1. *Initialization of the golden section method*

In practice, we choose the points x_1 and x_2 in the following way by taking $[a,b]=[0,1]$.

To illustrate: we set $ac = A$; $ad = B$; $ab = 1$ and write:

$$\frac{A}{B} = \frac{B}{1} \Rightarrow A = B^2$$

and $\frac{1-B}{1-A} = \frac{1-A}{1} \Rightarrow B^2 + B - 1 = 0$,

Now, the solution to equation $B^2 + B - 1 = 0$ is the golden ratio:

$$B = r = \frac{\sqrt{5}-1}{2} \quad A = 1 - r.$$

Our algorithm continues in this manner until the minimum of the function f is reached.

A11.3.2. *Descent methods*

A11.3.2.1. *Principle of descent methods*

Let $f: \mathbb{R}^n \to \mathbb{R}$ be a given differentiable function. Let $x \in \mathbb{R}^n$. The gradient vector $\nabla f(x)$ of f in x is a vector of $\mathbb{R}^n$., normal to the "surface" of the iso-values (recall that the sets of iso-values of f are defined, for $\alpha \in \mathbb{R}$ by $C_\alpha = \{x \in \mathbb{R}^n \,/\, f(x) = \alpha\}$) of f passing through x. It is also directed in the direction of the increasing values of f.

DEFINITION.– Direction of descent.

$d \in \mathbb{R}^n$ is a direction of descent for f at point x if:

$$\exists \tilde{t} > 0 \,/\, \forall t \in]0, \tilde{t}[,\ f(x + td) < f(x).$$

EXAMPLE.– Let x be such that $\nabla f(x) \neq 0$. Let a direction $d \in \mathbb{R}^n$ be such that:

$$\nabla f(x).d < 0.$$

So, d is a direction of descent.

REMARK A11.5.– It can be thus said that the directions that form an acute angle with the opposite of the gradient are descent directions. In fact, $\nabla f(x).d < 0 \leftrightarrow -\nabla f(x).d > 0 \leftrightarrow \theta$ acute.

where θ is defined by $\cos(\theta) = \dfrac{-\nabla f(x).d}{\| \nabla f(x).d \|}$.

Descent methods are the iterative methods. Their general scheme is as follows:

– $x^{(0)} \in \mathbb{R}^n$ given

– A sequence $x^{(1)}, \ldots, x^{(k)}$ is constructed by recurrence:

$$x^{(k+1)} = x^{(k)} + t_k d^k$$

where d^k is a direction of descent at point $x^{(k)}$ and $t_k > 0$ is the step of the method at iteration k.

Choice of step:

The parameter t_k must be chosen based on the behavior of the function of a real variable:

$$\phi(t) = f(x^{(k)} + td^k)$$

For a direction of descent d^k, we know that $\phi(0) = f(x^{(k)})$ and that $\phi'(0) = \nabla f(x^{(k)}).d^k < 0$.

The optimal choice is then to choose t_k such that:

$$\forall t > 0, \quad \phi(t_k) \leq \phi(t) \qquad \text{[A11.3]}$$

The descent method is then called *optimal step* descent method.

However, calculating the optimal step is a unidirectional optimization problem that is usually hard to solve.

A11.3.2.2. *The greatest slope method*

THEOREM.– Let $f: \mathbb{R}^n \to \mathbb{R}$ be a continuously differentiable function. Given $x \in \mathbb{R}^n$. In the vicinity of x, the direction in which f decreases the fastest is $d = -\nabla f(x)$.

– The steepest slope methods (or gradient methods) are generally found as:

$$\begin{cases} x^{(0)} \text{ given,} \\ x^{(1)},\ldots,x^{(k)} \text{ known :} \\ d^k = -\nabla f(x^{(k)})^T \\ x^{(k+1)} = x^{(k)} + t_k d^k \end{cases}$$

– The optimal step gradient method includes an optimal choice of the step t_k, as we saw previously. Then:

$$\forall t \geq 0,\ f(x^{(k+1)}) = f(x^{(k)} + td^k)$$

This step, called linear search, is in the case of any function f, a complex iterative process.

REMARK A11.6.– For an optimal-step gradient method, according to inequality [A11.3], it can be written that $\phi'(t_k) = 0$ (necessary condition of optimality). However:

$$\phi'(t_k) = \nabla f(x^{(k)} + t_k d^k).d^k$$

Therefore:

$$\nabla f(x^{(k+1)}).d^k = 0.$$

and two successive descent directions are orthogonal.

A11.4. Quadratic form

Let $A \in \mathcal{M}_{nn}(\mathbb{R})$ be a positive definite symmetric matrix. Let b be a vector of $\mathbb{R}^n$. Consider the quadratic function defined by:

$$f(x) = \frac{1}{2}x^T Ax - b^T x$$

Let d be any direction of $\mathbb{R}^n$ ($d \neq 0$). Therefore, for any $t \in \mathbb{R}$:

$$f(x+td) = \frac{1}{2}(d^T Ad)t^2 + d^T(Ax-b)t + \left(\frac{1}{2}x^T Ax - b^T x\right)$$

A being a positive definite matrix, it follows that $d^T Ad > 0$. Therefore, the function $t \mapsto f(x+td)$ is a parabola. The minimum is thus reached for:

$$\tilde{t} = -\frac{d^T(Ax-b)}{d^T Ad} \qquad \text{[A11.4]}$$

A11.4.1. *Optimal step gradient method*

We consider here the optimal step gradient method. Consequently, the descent direction is $d = -\nabla f(x)^T = Ax - b,$ from which:

$$\tilde{t} = \frac{\| d \|^2}{d^T Ad}$$

and the method in this case is written as:

Let $x^{(0)}$ be an initial piece of data; we calculate $d^{(0)} = b - Ax^{(0)}$, then for $k \geq 0$:

$$d^{(k)} = b - Ax^{(k)},$$

$$t_k = \frac{< d^{(k)}, d^{(k)} >}{< Ad^{(k)}, d^{(k)} >}$$

$$x^{(k+1)} = x^{(k)} + t_k d^{(k)}$$

A11.4.2. *Conjugate gradient method*

An even faster method in situation when A is a positive definite symmetric is that of the conjugate gradient. This time let:

$$x^{(k+1)} = x^{(k)} + \alpha_k d^{(k)}, k \geq 0, \qquad [A11.5]$$

with $d^{(k)} = r^{(k)} - \beta_k d^{(k-1)}$.

Let $x^{(0)}$ be an initial condition; we calculate $r^{(0)} = b - Ax^{(0)}$, $d^{(0)} = r^{(0)}$, then for $k \geq 0$:

$$\alpha_k = \frac{< r^{(k)}, d^{(k)} >}{< Ad^{(k)}, d^{(k)} >}$$

$$x^{(k+1)} = x^{(k)} + \alpha_k d^{(k)}$$

$$r^{(k+1)} = r^{(k)} - \alpha_k A d^{(k)}$$

$$\beta_{k+1} = \frac{< Ad^{(k)}, r^{(k+1)} >}{< Ad^{(k)}, d^{(k)} >}$$

$$d^{(k+1)} = r^{(k+1)} - \beta_k d^{(k)}.$$

A11.5. Optimization with constraints

The focus is now on the optimization problem with inequality constraints only:

$$(PCI)\begin{cases} \min_{x\in\mathbb{R}^n} & f(x), \\ \text{s. c.} & g(x) \leq 0, \end{cases}$$

where $g: \mathbb{R}^n \to \mathbb{R}^m$ is differentiable. We will denote by $.K.$ the set of the admissible points, that is:

$$K = \{x \in \mathbb{R}^n, g(x) \leq 0\}.$$

At the solution point of (PCI), it is obvious that the effectively active constraints will verify $g_i(\hat{x}) = 0$. However, since it is not previously known what these constraints are, shifting from (PCI) to a problem with equality constraints cannot be done directly.

The following results are first introduced:

THEOREM.– *Existence with constraints*

Let $K \subset \mathbb{R}^n$ be a closed part and $f: K \to \mathbb{R}$ a continuous function. In addition, we assume that:

$$\begin{cases} K \text{ limited} \\ or \\ f(x) \to +\infty \text{ when } \|x\| \to \infty, x \in K \end{cases}$$

Then, there exists at least one element $\overline{x} \in \mathbb{R}$ such that:

$$f(\overline{x}) \leq f(x) \quad \forall x \in K.$$

THEOREM.– *Uniqueness.*

Let K be a convex set of $\mathbb{R}^n$. Let $f: K \to \mathbb{R}\leq$ be a strictly convex function. As such, the minimum $\overline{x} \in K$, if it exists, is unique.

DEFINITION.– We call saturated constraints in $\hat{x}$ the set of indexes i such that $g_i(\hat{x}) = 0$, and we denote $I(\hat{x}) = \{i \mid g_i(\hat{x}) = 0\}$. We then denote by $S(\hat{x})$ the set $S(\hat{x}) = \{x \in \mathbb{R}^n, g_i(x) = 0, i \in I(\hat{x})\}$.

The concept of admissible direction is defined as follows:

DEFINITION.– $y \in \mathbb{R}^n$ is said to be an admissible direction in $\hat{x} \in K$ if there exists $\alpha > 0$ and a curve $x(t)$ verifying:

$$\begin{cases} x(t) & \in & K, \forall t \in [-\alpha, \alpha], \\ x(0) & = & \hat{x}, \\ x^{\cdot}(0) & = & y. \end{cases}$$

We will then denote $y \in C(\hat{x})$.

We have the following result:

Let $y \in \mathbb{R}^n$ be an admissible direction in $\hat{x} \in K$, then we necessarily have:

$$\nabla g_i(\hat{x})^T y \leq 0, i \in I(\hat{x}).$$

DEFINITION.– $\hat{x}$ is said to be a regular point for the constraint $g(x) \leq 0$ if:

– $g(\hat{x}) \leq 0$;

– the vectors $\{\nabla h_i(\hat{x})\}_{i \in I(\hat{x})}$ are linearly independent.

Under the assumption of regularity of $\hat{x}$, it will follow that, as for the equality constraints:

$$\nabla g_i(\hat{x})^T y \leq 0, i \in I(\hat{x})) \Rightarrow y \in C(\hat{x}).$$

The following proposition can be understood as the first step toward the Kuhn and Tucker conditions.

PROPOSAL.– Let $\hat{x}$ be the solution of the problem (PCI). There exists $\eta > 0$ such that $\forall x \in B(\hat{x},\eta), g_i(x) < 0, i \notin I(\hat{x})$, where we have denoted by $B(\hat{x},\eta)$. the ball of center $\hat{x}$ and radius η. Then, $\hat{x}$ is the solution to the problem:

$$\begin{cases} \min_{x\in B(\hat{x},\eta)} & f(x), \\ \quad \text{s.c.} & g_i(x) = 0, i \in I(\hat{x}). \end{cases}$$

This result is only due to the continuity of g and shows that the problem is locally reduced to a problem with equality constraints. The main result can now be stated:

THEOREM.– Let $\hat{x} \in K$ be a regular point solution to the problem (PCI). Then, there exists a single vector $\lambda \in \mathbb{R}^m$ such that:

$$\begin{aligned} \nabla f(\hat{x}) + \sum_{i=1}^{m} \nabla g_i(\hat{x}) &= 0, \\ \lambda_i &\geq 0, i = 1,\ldots,m, \\ \lambda_i g_i(\hat{x}) &= 0, i = 1,\ldots,m \end{aligned}$$

Appendix 12

Notion of Statistics

A12.1. Introduction

Statistics is a set of methods for collecting "raw" data, presenting and summarizing this data, and finally drawing conclusions about the population studied (its structure, its composition) to help in decision-making in the presence of time-dependent data and then to make forecasts.

Descriptive statistics addresses the second of these objectives. Many software programs are capable of ensuring the representation and processing aspects, while the specialist is left with the task of analyzing the results. First, the basic vocabulary of descriptive statistics will be introduced. We will then see how a statistical study can be represented as a table or graphs. Finally, the parameters most frequently used to summarize a quantitative statistical series are defined, depending on whether it involves a single or double series.

A12.2. Vocabulary

The population is a set comprising well-defined elements, with individuals that will be used as objects of study. In general, the population is too large to enable the study of all individuals. As such, observations are made with a subset of the population called a sample. It is hoped that this limited presentation will be able to draw "conclusions" about the entire population. The choice of this sample can be made in two ways: either it is constructed as a representation of the entire population, which presupposes a perfect knowledge thereof, or it is built by randomly searching the population. After this introduction of chance into the procedure, we rely on the theory of probabilities. As regard descriptive statistics, we merely consider the observations and try to "make them speak" as best as possible, without many assumptions about their origin.

For each individual in the population or sample, one or more aspect, called traits (or statistical variables), can be identified, which can be quantitative or qualitative. Quantitative traits are themselves split into two classes: discrete traits and continuous traits, depending on whether the values taken by these traits belong to discrete or continuous sets. For a discrete or qualitative trait, the values taken by the statistical variable (depending on the individual) are called modalities. For a continuous trait, the "possible values" are no longer in countable quantity. For practical reasons, this trait can be discretized by achieving a grouping in classes. A finite partition of the range of possible values is made; each element of the partition is called a class.

A12.3. Data tables

A12.3.1. *Person/trait table*

The individual/trait table is the matrix expression of the raw data, that is of the p traits recorded from the n population individuals or the sample (p and n being the positive integers). For $i \in \{1,\ldots,n\}$, $j \in \{1,\ldots,p\}$, $X_{i,j}$ denotes the "value" (quantitative or qualitative) of the j-th for the i-th individual. The raw data is then presented in a two-way table whose element in the i-th row and j-th column is $X_{i,j}$ or the transposed table.

The one-way or single statistical series table provides a more compact representation of the observations when only one trait is observed. The frequency of a class is the ratio of the number of elements in the class to the total number of students. A representation of the series as a table including the class and number of elements or a table based on class and frequency or both in one. This will then be referred to as the distribution of the statistical variable.

Formally, let X be the trait under study and it is assumed that it is quantitative discrete, taking the values $x_1,x_2,\ldots,x_q$. The sample size is n, the measured values are $X_1,\ldots,X_n$, among which n_i times the value x_i. $f_i = \frac{n_i}{n}$. denotes the associated frequencies. A table is obtained as:

x_i	x_1	$\ldots$	x_q
n_i	n_1	$\ldots$	n_q
f_i	f_1	$\ldots$	f_q

It can be observed that $\sum_{i=1}^{q} n_i = n$ and $\sum_{i=1}^{q} f_i = 1$. A statistical trait represented in this way is called a statistical series.

Contingency tables are the common name given to two-way tables for simultaneously representing two traits. These are assumed to be qualitative or discrete, or continuous, but in this case, they are grouped in classes. The raw series (called double series) is then (X_1, Y_1), $\cdots$, (X_n, Y_n). If there are p classes $x_1, \ldots, x_p$ for X and q classes $y_1, \ldots, y_q$ for Y, there are $p \times q$ possible values for each pair (X_i, Y_i); these values are all the (x_i, y_j). The number of occurrences of each pair (x_i, y_j) is counted, and we denote this number by $n_{i,j}$. The frequency of the pair is denoted by $f_{i,j} = \frac{n_{i,j}}{n}$. The contingency table is the matrix $(n_{i,j})_{i=1,\ldots,p,j=1,\ldots,q}$ or $(f_{i,j})_{i=1,\ldots,p,j=1,\ldots,q}$.

The marginal numbers are obtained by simply summing the numbers according to the rows (row and column). These distributions are thus said to be marginal, as opposed to the joint distribution that can be read in the table.

The purpose of the joint distribution is the value of a pair of statistical variables. The respective objects of the marginal distributions are the two variables taken separately. Let us assume that we are interested in the distribution of X assuming that Y has any value, that is, y_0. We then only consider a subset of the population, the one composed of individuals whose trait Y is equal to y_0. Then, the distribution of X conditionally to $Y = y_0$, or still the conditional distribution of X knowing that $Y = y_0$ is given by the row of the table $Y = y_0$.

The conditional frequency distribution is obtained by dividing the conditional numbers by the marginal numbers $Y = y_0$ that can be read at the end of the row $Y = y_0$.

A12.3.2. *Chi-squared distance*

Observing one (or two) trait(s) in two different populations, two different distributions are obtained a priori as two simple statistical series or two contingency tables. To decide whether the distributions obtained are substantially the same and to conclude that in view of the measured trait(s), the populations are the same, we have the chi-squared distance (or χ^2) which is written as follows, for two f_i and f_i' frequency tables:

$$d_{\chi^2} = \sum_i \frac{(fi - f_i')^2}{f_i}.$$

In this sum, i is the column index in the one-dimensional case or the double row-column index in the two-dimensional case. This (positive) distance increases when the difference between the frequencies observed increases.

A12.4. Graphical representations

A12.4.1. *Representation of the values*

A qualitative trait is often represented by a circular diagram (pie chart), where the angular sector of each modality is proportional to the number of individuals in the modality, with a 360° angle indicating the total number (see Figure A12.1).

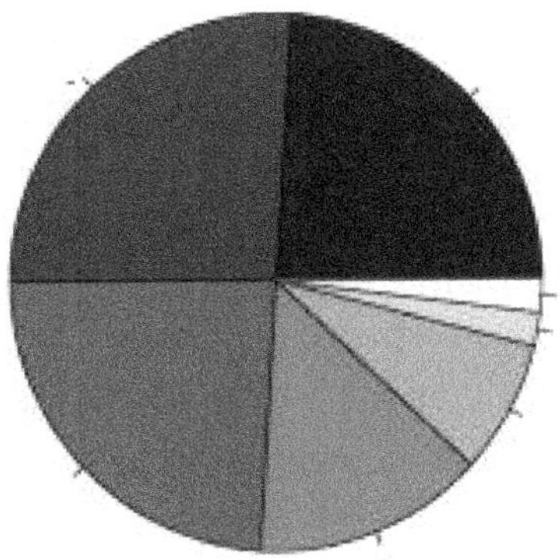

Figure A12.1. *Pie chart*

Discrete quantitative traits are graphically represented by bar charts (see Figure A12.2).

For continuous quantitative characteristics, a grouping by class must be carried out before drawing a histogram (see Figure A12.3).

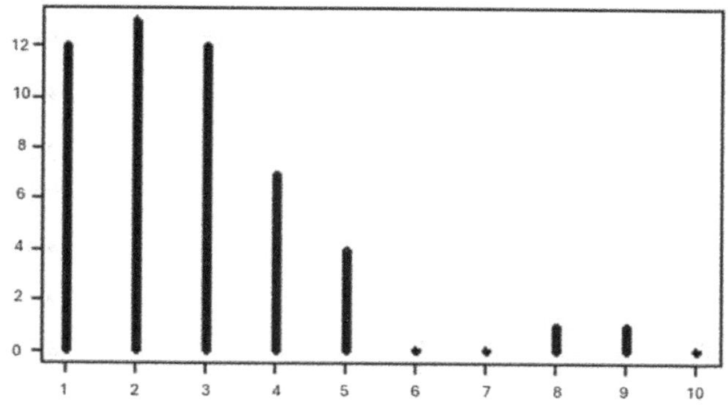

Figure A12.2. *Bar chart*

A12.4.2. *Cumulative values*

This question concerns only quantitative traits.

DEFINITION.– In a statistical series of length n relating to a given trait, the function of the cumulative figures is the function $H_n(x)$ which, with any value x associates the number of observations less than or equal to x.

The function of the cumulative frequencies $F_n(x)$ associates with any value x the proportion of observations less than x with respect to the total number of elements. These functions are positive, increasing from 0 to n and related by the relation $F_n(x) = \frac{1}{n} H_n(x)$.

In situations where the variable is discrete, the cumulative frequency function looks like a "staircase" (see Figure A12.4).

The points of discontinuity are the modalities of the series, and the height of the steps at these points is equal to the number of values in the modalities in question.

When the variable is continuous, it is assumed that the value of $H_n(x)$. (and $F_n(x)$) is known at the boundaries of the classes, we place the $H_n(x_i)$ and link them by linear interpolation: this is then referred to as a cumulative frequency polygon (see Figure A12.5).

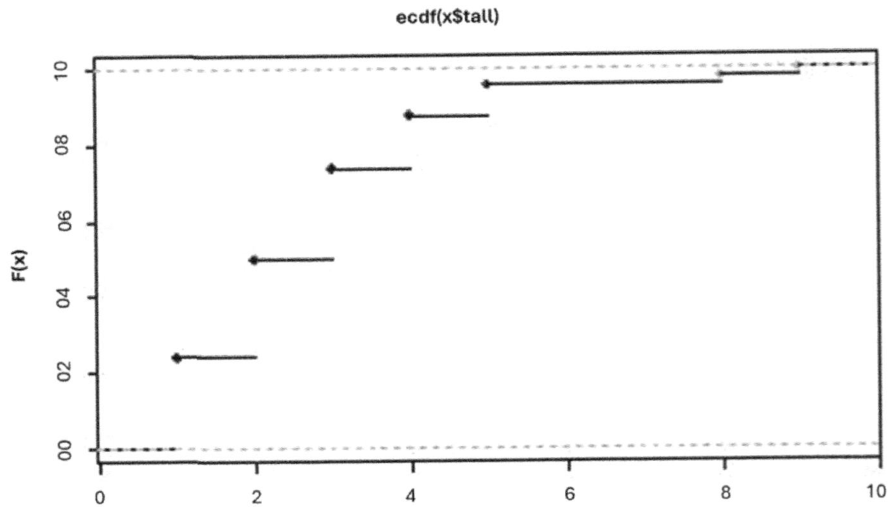

Figure A12.3. *Cumulative frequencies for the discrete trait*

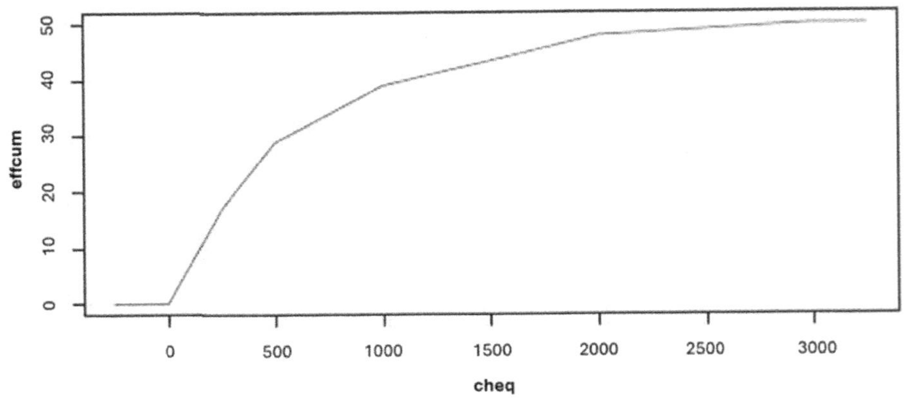

Figure A12.4. *Cumulative values polygon*

A12.5. Parameters of a simple series

The parameters of a statistical series are numerical quantities that are supposed to "summarize" the series or reflect a certain aspect of it in a compact way (including almost always loss of information).

A distinction is made between position (or central tendency) and dispersion parameters. When a statistical series is translated (a fixed quantity is added to each observation), the position parameters (such as the mean or the median) are translated by the same amount. The dispersion parameters remain unchanged. They measure the spread of data (concentrated or dispersed) relatively to the central tendency. There are other parameters, such as those measuring the skewness of the distribution.

The minimum and maximum values x_{min} and x_{max} of the observations give an idea, obviously insufficient, of the area where they are located. These are position parameters. The extent of the series $E = x_{max} - x_{min}$ is a dispersion parameter.

The mean of a statistical series is simply the arithmetic average (for a quantitative trait of course) of the observations. If these are n observations, denoted by $X_1, \ldots, X_n$, the average is defined and denoted by:

$$\overline{X} = \frac{1}{n}\sum_{i=1}^{n} X_i$$

If the series is available following a modality/effective scheme, the average is also given by the formula:

$$\overline{X} = \frac{1}{n}\sum_{j=1}^{p} n_j x_j$$

when the modalities are the values x_j and their respective individuals the n_j. When a statistical series by class is available without the raw data, only an approximate value can be given by assigning to each observation in an interval the average value of that observation (the middle of the interval).

DEFINITION.– The median of a statistical variable refers to any quantity M such that at least half of the observations are less than M and at least half of the observations are greater than M.

There is no uniqueness of the median in all generality. The set of medians is a closed interval eventually reduced to a singleton. A median is more sensitive to data close to the central tendency of the statistical variable and less to distant values. It is then preferred to the mean to determine the central tendency when there is suspicion that the series is corrupted by outliers. The three quartiles (the second being the median) and generally speaking the quantiles are defined.

The quantile function $Q(x)$ associates a proportion x (between 0 and 1) with a limit value (or interval) that leaves at least one proportion x of observations below it and at least $1-x$ above.

To determine the quantiles "by hand", the raw series is written in an orderly manner and split in half or four. Exact values or intervals are then obtained. To graphically determine the quantiles, for example, the quantile α, the curve (staircase or polygon) of the cumulative frequencies is made to intercept the line $y=\alpha$ ($\alpha=0.5$ to find the median).

The variance of the raw data $X_1,\ldots,X_n$ is denoted and defined by

$$\mathrm{var}(X)=\frac{1}{n}\sum_{i=1}^{n}(X_i-\bar{X})^2.$$

Normalization can sometimes be $n-1$. Nothing changes much when n is large. From this formula, it can be seen that the variance is small if all of the observations are close to their mean. The standard deviation is defined as the square root of this positive quantity:

$$\sigma_X=\sqrt{var(X)}.$$

The standard deviation is homogeneous to the physical dimension of the data, if they have one. Variance and standard deviation are dispersion parameters. If all of the observations are multiplied by a non-zero constant a, the variance is multiplied by a^2 and the standard deviation by $|a|$.

PROPOSITION (KOENIG).– for all quantitative data X, we have var $(X)=\overline{X^2}-(\bar{X})^2$.

If the series is known following a modality/number of individual schemes, another expression is obtained for the variance:

$$var(X)=\frac{1}{n}\sum_{j=1}^{p}n_j(x_j-\bar{X})^2.$$

A12.6. Representations and parameters of a double series and linear regression

It is assumed that we want to study two statistical variables X and Y. Naturally, these can be addressed as two simple series and the means X and Y, the standard deviations σ_X and σ_Y can be calculated but it results in losing the whole link between X and Y. So, how can this link be "measured"?

The joint distribution can be represented as a mosaic or bar chart. We start by representing the point cloud $M_i(X_i, Y_i)$ in a Cartesian coordinate system (see Figure A12.6).

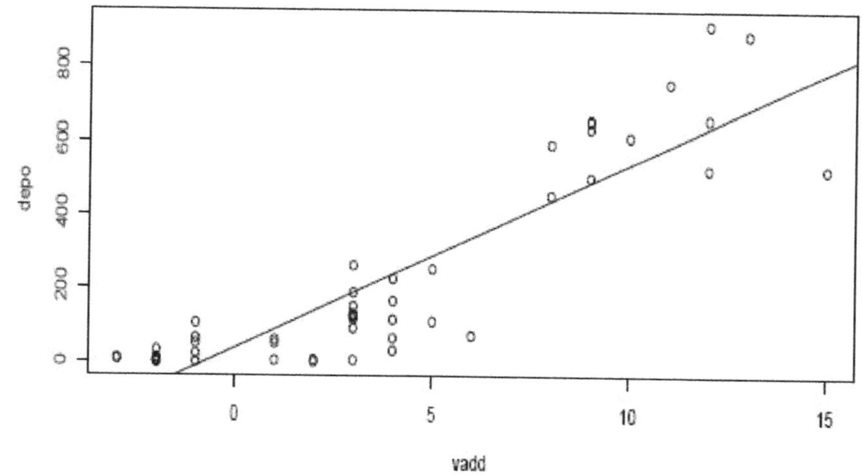

Figure A12.5. *A scatter plot*

The long-term objective is to decide whether there is a strong functional dependency between the two statistical parameters, that is, if we approximately have $Y = f(X)$.

If this is the case, the collection of points is concentrated around the curve representing the function *f*, and it becomes possible to control the second if we know how to set the first, that is, to predict the second if we only observe the first. These two operations are all the more accurate, the stronger the dependency.

First step: classification

We start by looking at whether the cloud is not composed of groups or families of points, in well-separated areas of the plane; this would correspond to very different regimes in the possible relation $Y = f(X)$. If this is the case, there can be a tendency to make a classification, or in other words, to label the observations.

Second step: regression

If the collection of points seems to concentrate around a certain curve of the equation $y = f(x)$, it becomes desirable to determine this function *f*. We must be able to quantify the accuracy of the statistical modeling $Y = f(X)$ for any function f. and then find the f. optimal function. The solution to this problem is particularly simple when looking for an affine function f., and when the criterion used to evaluate the approximation $Y = f(X)$ is the least squares criterion. This is called a linear regression of Y on X.

DEFINITION.– We call covariance of X. and Y the number:

$$\operatorname{cov}(X,Y) = \operatorname{cov}(Y,X) = \frac{1}{n}\sum_{i=1}^{n}(X_i - \bar{X})(Y_i - \bar{Y})$$

and the correlation coefficient between X. and Y the number:

$$\rho_{X,Y} = \frac{\operatorname{cov}(X,Y)}{\sigma_X \sigma_Y}.$$

We have the following results:

– $\operatorname{cov}(X,Y) = \overline{X.Y} - \bar{X}.\bar{Y}$;

– $|\rho_{X,Y}| \leq 1$.

The linear regression of Y with respect to X makes it possible to look for the linear dependency relation $Y = aX + b$ that best summarizes the observations. The coefficients a and b of this regression are obtained by minimizing a “least squares” criterion. We look for the values a and b that minimize the sum of the squared errors that we make by approximating Y by $aX+b$. This criterion, as a function of two variables, is written as:

$$C(a,b)=\sum_{i=1}^{n}(Y_i-(aX_i+b))^2.$$

This criterion can be minimized by writing that (a,b) is an extremum provided that:

$$\frac{\partial C(a,b)}{\partial a}=\frac{\partial C(a,b)}{\partial b}=0.$$

Solving this system leads to the following solution:

$$\hat{a}=\frac{\operatorname{cov}(X,Y)}{\sigma_X^2} \text{ and } \hat{b}=\overline{Y}-\hat{a}\overline{X}.$$

By studying the second derivative of the criterion, it is easy to see that it is indeed a minimum and not a maximum. The equation of the regression line of Y with respect to X is $y=\hat{a}x+\hat{b}$, or still:

$$y-\overline{Y}=\frac{\operatorname{cov}(X,Y)}{\sigma_X^2}(x-\overline{X})$$

while that of X with respect to Y is symmetrically written as:

$$x-\overline{X}=\frac{\operatorname{cov}(X,Y)}{\sigma_Y^2}(y-\overline{Y}).$$

A12.7. Estimates

Consider a population P with N the number of individuals. It is assumed that, for the observed trait, the mean of P is m while its standard deviation is σ. These are the two values that we want to find from the samples. It is therefore assumed that there are k samples of P, each of which has n individuals. We denote by E_1, E_2, ..., E_k these k samples, of respective means $\overline{x_1}$, $\overline{x_2}$, ..., $\overline{x_k}$, and of respective standard deviations σ_1, σ_2, ..., σ_k. The set $\overline{X}=\{\overline{x_1},\overline{x_2},\ldots,\overline{x_k}\}$ is a statistical series of k values, this series is called the distribution of the mean. It is then shown that:

$$E(\overline{X}) = m \quad \text{and} \quad \sigma_{\overline{X}} = \tfrac{\sigma}{\sqrt{n}}.$$

Moreover, for $n > 30$, the random variable $\overline{X}$ follows approximately a normal distribution $\mathcal{N}(m, \dfrac{\sigma}{\sqrt{n}})$. In other words, the random variable $\dfrac{\overline{X} - m}{\sigma / \sqrt{n}}$ follows approximately a normal distribution $\mathcal{N}(0,1)$.

A12.7.1. *Point estimate*

DESCRIPTION.–

– Knowing the mean x and standard deviation σ' of a sample of size n, the goal is to estimate the mean m, the standard deviation σ, and the variance σ^2 of the total population. For the mean, point estimation is the method that consists of equating the measurement with the sample with the mean of the total population. It will be said that x is a point estimate of the mean m. For the variance and standard deviation, it is assumed that: $\dfrac{n}{n-1}\sigma'^2$ is a point estimate of the variance σ^2 and therefore: $\sqrt{\dfrac{n}{n-1}}\sigma'$ is a point estimate of the standard deviation σ.

– If, for a full population, it is a frequency p that is to be estimated from the frequency f observed from a sample, the procedure is the same as for a mean. More specifically, point estimation is the method that consists of equating the measurement with the sample with the measurement of the total population. If f is the frequency, in the sample, of the observed characteristic, it will be said that f is a point estimate of the frequency p.

A12.7.2. *Mean estimation for a confidence interval*

Consider a population P with N the number of individuals. It is assumed that, for the trait being observed, the unknown mean of P is m while its known standard deviation is σ. A succession of samples of the same size is drawn at random, and with bootstrapping, the n respective averages of which are calculated: x_1 for the first, x_2 for the second, etc.

X denotes the random variable that associates a sample E_i with its mean x_i. The variable X therefore successively takes the values $\overline{x_1}$, $\overline{x_2}$, etc. Finally, it is also assumed that the conditions are met to be able to use a consequence of the centered limit theorem and make the approximation that X follows the normal distribution $\mathcal{N}(m, \sigma / \sqrt{n})$. In other words, that the random variable $T = \frac{\sqrt{n}}{\sigma}(\overline{X} - m)$ follows the normal distribution $\mathcal{N}(0,1)$. We will then get, for all $t \geq 0$,

$$Pr[-t \leq T \leq t] = 2\Phi(t) - 1.$$

A12.7.3. *Frequency estimation for a confidence interval*

Consider a population P with N the number of individuals. It is assumed that, for the observed trait, the unknown frequency of P is p. It is also assumed that we have a sample of n of this population, a sample in which the observed trait is observed with the frequency f.

We are now at the point where we can consider that the variable F which, for any random sample of fixed size n, associates the frequency of this sample exhibiting the fixed property, follows the normal distribution $\mathcal{N}(p, \sqrt{p(1-p)/n})$. A reasoning and computations similar to those in the previous paragraph give the following result: the interval:

$$\left[f - t\sqrt{\frac{f(1-f)}{n-1}}, f + t\sqrt{\frac{f(1-f)}{n-1}} \right]$$

is the confidence interval of the frequency p of the population with the confidence coefficient $2\Phi(t) - 1$, centered on the frequency f of the sample under consideration. In practice, confidence coefficients of 95% are often employed, which corresponds to $t = 1.96$, or 99%, which corresponds to $t = 2{,}58$.

References

[ABI 19] ABID F., Contribution à la robustesse et à l'optimisation fiabiliste des structures, Thesis, INSA Rouen Normandie, defended 2 December 2019.

[ABI 20] ABID F., DAMMAK K., EL HAMI A. et al., "Surrogate models for uncertainty analysis of micro-actuator", *Microsystem Technologies*, pp. 1–12, 2020.

[ABO 11a] ABO AL-KHEER A., EID M., AOUES Y. et al., "Theoretical analysis of the spatial variability in tillage forces for fatigue analysis of tillage machines", *Journal of Terramechanics*, vol. 48, no. 4, pp. 285–295, 2011.

[ABO 11b] ABO AL-KHEER A., KHARMANDA M.G., EL HAMI A. et al., "Estimating the variability of tillage forces on a chisel plough shank by modeling the variability of tillage system parameters", *Computers and Electronics in Agriculture*, vol. 78, no. 1, pp. 61–70, 2011.

[ABO 11c] ABO AL-KHEER A., EL-HAMI A., KHARMANDA M.G. et al., "Reliability-based design for soil tillage machines", *Journal of Terramechanics*, vol. 48, no. 1, pp. 57–64, 2011.

[AFO 02] AFONSO S.M.B., MACEDO C.M.H., OLIVEIRA D.A.P., "Structural shape optimization under multicriteria conditions", in *WCCMV – Fifth World Congress on Computational Mechanics*, 7–12 July 2002.

[ALL 01] ALLAIRE G., HENROT, A., "On some recent advances in shape optimization", *Comptes Rendus de l'Academie des Sciences, Série II b*, vol. 329, pp. 383–396, 2001.

[ALR 12] ALRAJEH N.A., ELMIR B., BOUNABAT B. et al., "Interoperability optimization in healthcare collaboration networks", *Biomedizinische Technik*, vol. 57, no. 5, pp. 403–411, 2012.

[AMM 14] AMMARI N., GHALLALI M., EL KALAM A.A. et al., "Mobile security: Security mechanisms and protection of mobile applications", *Journal of Theoretical and Applied Information Technology*, vol. 70, no. 2, pp. 302–315, 2014.

[ASN 03] ASNAFI N., "Analytical modelling of tube hydroforming", *Thin-Walled Structures*, vol. 34, pp. 295–330, 2003.

[AYA 11] AYADI A. RADI B., CHEROUAT A., EL HAMI A. "Optimization and identification of the characteristics of a hydroformed structures", *Applied Mechanics and Materials*, pp. 11–20, 2011.

[AYY 97] AYYUB B., MCUEN, R.H., *Probability, Statistics & Reliability for Engineers*, CRC Press, Boca Raton, FL, 1997.

[AZI 02] AZID I.A., KWAN A.S.K., SEETHARAMU K.N., "A GA-based technique for layout optimization of truss with stress and displacement constraints", *International Journal for Numerical Methods in Engineering*, vol. 53, pp. 641–1674, 2002.

[BAR 99] BARAKAT A.S., MALKAWI A.I.H., TAHAT R.H. "Reliability based design optimization of laterally loaded piles", *Stuctural Safety*, vol. 7, pp. 45–64, 1999.

[BEN 90] BEN-HAIM Y., ELISHAKOFF I., *Convex Models of Uncertainty in Applied Mechanics*, Elsevier Science, Amsterdam, 1990.

[BEN 08] BENASCIUTTI D., CRISTOFORI A., "A frequency-domain formulation of MCE method for multiaxial random loading", *Fatigue & Fracture of Engineering Materials & Structures*, vol. 31, pp. 937–948, 2008.

[BEN 11] BEN ABDESSALEM M.A., Optimisation avec prise en compte des incertitudes dans la mise en forme par hydroformage, Thesis, INSA Rouen Normandie, defended 8 June 2011.

[BEN 13] BEN ABDESSALEM A., PAGNACCO E., EL-HAMI A., "Increasing the stability of T-shape tube hydroforming process under stochastic framework", *International Journal of Advanced Manufacturing Technology*, vol. 69, nos 5–8, pp. 1343–1357, 2013.

[BEN 14] BEN ABDESSALEM A., EL HAMI A., "Global sensitivity analysis and multi-objective optimisation of loading path in tube hydroforming process based on metamodelling techniques", *International Journal of Advanced Manufacturing Technology*, vol. 71, nos 5–8, pp. 753–773, 2014.

[BEN 15] BEN ABDESSALEM A., EL HAMI A., "A probabilistic approach for optimising hydroformed structures using local surrogate models to control failures", *International Journal of Mechanical Sciences*, vols 96–97, pp. 143–162, 2015.

[BOU 96] BOUYEKHF R., EL MOUDNI A., EL HAMI A., "Reduced order modelling of two-time-scale discrete non-linear systems", *Journal of the Franklin Institute*, vol. 333, no. 4, pp. 499–512, 1996.

[BOU 01a] BOUCHEKOURTE M., RHOUAS S., EL HAMI, N., "Minimizing the variance of the coverage ratio as an approach to optimize the exchange rate risk of Brent futures contracts", *International Journal for Simulation and Multidisciplinary Design Optimization*, vol. 2022, no. 13, p. 17, 2001.

[BOU 01b] BOUYEKHF R., EL HAMI A., EL MOUDNI A., "Optimal control of a particular class of singularly perturbed nonlinear discrete-time systems", *IEEE Transactions on Automatic Control*, vol. 46, no. 7, pp. 1097–1101, 2001.

[BRE 84] BREITUNG K., "Asymptotic approximations for multi-normal integral", *Journal of Engineering Mechanics*, vol. 110, pp. 357–366, 1984.

[BRU 07] BRUYERE J., DANTAN J.-Y., BIGOT R. et al., "Statistical tolerance analysis of bevel gear by tooth contact analysis and Monte Carlo simulation", *Mechanism and Machine Theory*, vol. 42, no. 10, pp. 1326–1351, 2007.

[BUC 90] BUCHER C.G., BOURGUND U., "Fast and efficient response surface approach for structural reliability problem", *Structural Safety*, vol. 7, pp. 57–66, 1990.

[CAM 00] CAMBIER S., Approche probabiliste pour la prise en compte de la dispersion de paramètres mécaniques : application à la fatigue vibratoire de réseaux de tuyauteries, Thesis, Ecole nationale supérieures d'arts et métiers, Paris, 2000.

[CHE 08] CHEROUAT A., RADI B., EL HAMI A., "The frictional contact of the composite fabric's shaping", *Acta Mechanica*, vol. 199, nos 1–4, pp. 29–41, 2008.

[COM 01] COMBESCURE A., GUSIC G.., "Nonlinear buckling of cylinders under external pressure with nonaxismmetric thickness imperfections using the COMI axisymmetric shell element", *International Journal of Solids and Structures*, vol. 38, pp. 6207–6226, 2001.

[COU 06] COUEGNAT G., FOK S.L., COOPER J.E. et al., "Structural optimization of dental restorations using the principle of adaptive growth", *Dental Materials*, vol. 22, no. 1, pp. 3–12, 2006.

[CUL 94] CULIOLI J.-C., *Introduction à l'optimisation*, Ellipses, Paris, 1994.

[DAM 17] DAMMAK K., EL HAMI A., KOUBAA S. et al., "Reliability based design optimization of coupled acoustic-structure system using generalized polynomial chaos", *International Journal of Mechanical Sciences*, vol. 134, pp. 75–84, 2017.

[DAM 19a] DAMMAK K., EL HAMI A., "Multi-objective reliability-based design optimization of coupled acoustic-structural system", *Engineering Structures*, vol. 197, p. 109389, 2019.

[DAM 19b] DAMMAK K., KOUBAA S., EL HAMI A. et al., "Numerical modelling of vibro-acoustic problem in presence of uncertainty: Application to a vehicle cabin", *Applied Acoustics*, vol. 144, pp. 113–123, 2019.

[DAM 20a] DAMMAK K., EL HAMI A., "Multi-objective reliability-based design optimization using Kriging surrogate model for cementless hip prosthesis", *Computer Methods in Biomechanics and Biomedical Engineering*, vol. 23, no. 12, pp. 854–865, 2020.

[DAM 20b] DAMMAK K., YAICH A., EL HAMI A. et al., "An efficient optimization based on the robust hybrid method for the coupled acoustic–structural system", *Mechanics of Advanced Materials and Structures*, vol. 27, no. 21, pp. 1816–1826, 2020.

[DAM 21] DAMMAK K., EL HAMI A., "Thermal reliability-based design optimization using Kriging model of PCM based pin fin heat sink", *International Journal of Heat and Mass Transfer*, vol. 166, p. 120745, 2021.

[DAV 64] DAVENPORT A.G., "Note on the distribution of largest values of random function with application to gust loading", *Proceedings of the Institution of Civil Engineers*, vol. 28, pp. 187–196, 1964.

[DEB 99] DEB K., Multi-objective evolutionary algorithms: Introducing bias among pareto-optimal solutions, KanGal Report 99002, Indian Institute of Technology, Kanpur, 1999.

[DEB 20] DEBICH B., EL HAMI A., YAICH A. et al., "An efficient reliability-based design optimization study for PCM-based heat-sink used for cooling electronic devices", *Mechanics of Advanced Materials and Structures*, pp. 1–13, 2020.

[DEB 21] DEBICH D., Modélisation, simulation, et fiabilité des matériaux changement de phase pour l'automobile et l'aéronautique, Thesis, INSA Rouen Normandie, 2021.

[DER 87] DER KIUREGHIAN A., LIN H.Z.S.J., "Second-order reliability approximations", *Journal of Engineering Mechanic*, vol. 113, pp. 1208–1225, 1987.

[DER 91] DER KIUREGHIAN A., DE STEFANO MARIO "Efficient algorithm for second-order reliability analysis", *Journal of Engineering Mechanics*, vol. 117, pp. 2904–2923, 1991.

[DES 00] DESSOMBZ O., Analyse dynamique de structures comportant des paramètres incertains, Thesis, Ecole centrale de Lyon, 2000.

[DUB 88] DUBOIS D., PRADE H., *Théorie des possibilités : application à la présentation des connaissances informatiques*, Masson, Paris, 1988.

[DUY 96] DUYSINX P., Optimisation topologique : du milieu continu a la structure élastique, Thesis, Université de Liége faculté des Science Appliquées, 1996.

[EDD 19] ED-DAOUI I., KOULOU A., HAMI N.E. et al., "An approach to systems-of-systems structural analysis through interoperability assessment: Application on Moroccan case", *International Journal of Engineering Research in Africa*, vol. 41, pp. 175–189, 2019.

[ELA 23] EL ATTAOUI A., BOUKHAFLA A., RHOUAS S. et al., "Custom application programming interface data extractor applied to the Klarna e-commerce dataset", *Indonesian Journal of Electrical Engineering and Computer Science*, vol. 31, no. 3, pp. 1624–1632, 2023.

[ELH 93] EL HAMI A., LALLEMENT G., MINOTTI P. et al., "Methods that combine finite group theory with component mode synthesis in the analysis of repetitive structures", *Computers and Structures*, vol. 48, no. 6, pp. 975–982, 1993.

[ELH 96] El HAMI A., RADI B., "Some decomposition methods in the analysis of repetitive structures", *Computers and Structures*, vol. 58, no. 5, pp. 973–980, 1996.

[ELH 09] El HAMI A., RADI B., CHEROUAT A., "Treatment of the composite fabric's shaping using a Lagrangian formulation", *Mathematical and Computer Modelling*, vol. 49, nos 7–8, pp. 1337–1349, 2009.

[ELH 11] El HAMI A., RADI B., "Comparison study of different reliability-based design optimization approaches", *Advanced Materials Research*, vol. 274, pp. 113–121, 2011.

[ELH 13] EL HAMI A., RADI B., *Uncertainty and Optimization in Structural Mechanics*, ISTE Ltd, London and John Wiley & Sons, New York, 2013.

[ELH 15] EL HAMI A., RADI B., "Analysis of the hydroforming process using metamodels", *Advances in Theoretical and Applied Mechanics*, pp. 52–60, 2015.

[ENE 93] ENEVOLDSEN I., SØRENSEN J.D., "Reliability-based optimisation of series systems of parallel systems", *Journal of Structural Engineering*, vol. 119, pp. 1069–1084, 1993.

[ENE 94] ENEVOLDSEN I., SØRENSEN J.D., "Reliability-based optimization in structural engineering", *Structural Safety*, vol. 15, pp. 169–196, 1994.

[FAN 03] FANN K.J., HSIAO P.Y., "Optimization of loading conditions for tube hydroforming", *Journal of Materials Processing Technology*, vol. 140, pp. 520–524, 2003.

[FEN 86] FENG Y.S., MOSES F., "Methods of structural optimization based on structural system reliability", *Journal of structural Mechanics*, vol. 14, pp. 437–453, 1986.

[FIE 79] FIESSLER B., NEUMANN H.J., RACKWITZ R., "Quadratic limit states in structural reliability", *Journal of the Engineering Mechanics Division*, vol. 105, pp. 661–676, 1979.

[FLE 87] FLETCHER R., *Practical Methods of Optimization*, John Wiley & Sons, New York, 1987.

[FRA 85] FRANGOPOL D.M. "Sensitivity of reliability-based optimum design", *Journal of Structural Engineering*, vol. 111, pp. 1703–1721, 1985.

[FRA 94] FRANGOPOL D.M., HENDAWI S., "Incorporation of corrosion effects in reliability based design optimisation of composite hybrid plate girders", *Structural Safety*, vol. 16, pp. 145–169, 1994.

[FU 90] FU G., FRANGOPOL D.M., "Reliability-based vector optimization of structural systems", *Journal of Structural Engineering*, vol. 116, pp. 2143–2161, 1990.

[GAS 97] GASSER M., SCHUELLER G.I., "Reliability-based optimization of structural systems", *Mathematical Methods of Operations Research*, vol. 46, pp. 287–307, 1997.

[GHA 91] GHANEM R.G., POL D., *Spanos Stochastic Finite Elements: A Spectral Approach*, Springer-Verlag, New York, 1991.

[GHA 96] GHANEM R.G., KRUGER R.M., "Numerical solution of spectral stochastic finite element systems", *Computer Methods in Applied Mechanics and Engineering*, vol. 129, pp. 289–203, 1996.

[GMÜ 97] GMÜR T., *Dynamique des structures*. Presses Polytechnique et Universitaires Romandes, 1997.

[GOL 83] GOLLWITZER RACKWITZ R., "Equivalent components in first-order system reliability", *Reliability Engineering*, vol. 5, pp. 99–115, 1983.

[GUE 15a] GUERINE A., EL HAMI A., WALHA L. et al., "Approach for the dynamic analysis of one stage gear system with uncertain parameters", *Mechanism and Machine Theory*, vol. 92, pp. 113–126, 2015.

[GUE 15b] GUERINE A., EL HAMI A., FAKHFAKH T. et al., "A polynomial chaos method to the analysis of the dynamic behavior of spur gear system", *Structural Engineering and Mechanics*, vol. 53, no. 4, pp. 819–831, 2015.

[GUE 16a] GUERINE A., EL HAMI A., WALHA L. et al., "A polynomial chaos method for the analysis of the dynamic behavior of uncertain gear friction system", *European Journal of Mechanics, A/Solids*, vol. 59, pp. 76–84, 2016.

[GUE 16b] GUERINE A., HAMI A.E., WALHA L. et al., "Dynamic response of a spur gear system with uncertain friction coefficient", *Advances in Engineering Software*, vol. 120, pp. 45–54, 2016.

[GUE 16c] GUERINE A., Contribution à l'étude du comportement dynamique d'un système d'engrenage en présence d'incertitudes, Thesis, INSA Rouen Normandie, defended 19 September 2016.

[GUE 17] GUERINE A., EL HAMI A., WALHA L. et al., "Dynamic response of wind turbine gear system with uncertain-but-bounded parameters using interval analysis method", *Renewable Energy*, vol. 113, pp. 679–687, 2017.

[HAM 64] HAMMERSLEY J., HANDSCOMB D., *Monte Carlo Methods*, Methuen, London, 1964.

[HAR 83] HARBITZ A., VERITAS D.N., "Efficient and accurate probability of failure calculation by use of the importance sampling", in *Fourth International Conference on Applications of Statistics and Probability in Soil and Structural Engineering*, Florence, pp. 825–836, 1983.

[HAS 74] HASOFER, A.M., LIND N., "An exact and invariant first-order reliability format", *Journal of Engineering Mechanics*, vol. 100, pp. 111–121, 1974.

[HED 11] HEDIA H.S., "Numerical optimization of dental implant through the concept of functionally graded materials", *Materials Testing*, vol. 53, no. 9, pp. 556–563, 2011.

[HOH 83] HOHENBICHLER M., Rackwitz R., "First-order in system reliability", *Structural Safety*, vol. 1, pp. 177–188, 1983.

[HOH 87] HOHENBICHLER M., GOLLWITZER S., KRUSE W. et al., "New light on first and second-order reliability methods", *Structural Safety*, pp. 267–284, 1987.

[HUA 16a] HUANG C., RADI B., EL HAMI A., “Uncertainty analysis of deep drawing using surrogate model based probabilistic method”, *The International Journal of Advanced Manufacturing Technology*, vol. 86, pp. 3229–3240, doi: 10.1007/s00170-016-8436-4, 2016.

[HUA 16b] HUANG C., RADI B., EL HAMI A., “Uncertainty analysis of deep drawing using surrogate model based probabilistic method”, *Manufacturing Technology*, vol. 86, nos 9–12, pp. 3229–3240, 2016.

[IBR 87] IBRAHIM R.A., “Structural dynamics with parameters uncertainties”, *Applied Mechanics Reviews*, vol. 40, no. 30, pp. 309–328, 1987.

[IMA 05] IMANINEJAD M., SUBHASH G., LOUKUS A., “Loading path optimization of tube hydroforming process”, *International Journal of Machine Tools and Manufacture*, vol. 45, pp. 1504–1515, 2005.

[IMB 95] IMBERT J.F., *Analyse des structures par éléments finis*, Cepadues, Toulouse, 1995.

[JAN 08] JANSSON T., NILSSON L., MOSHFEGH R., “Reliability analysis of a sheet metal forming process using Monte Carlo analysis and metamodels”, *Materials Processing Technology*, vol. 22, no. 1, 2008.

[JAN 17] JANNOUN M., AOUES Y., PAGNACCO E. et al., “Probabilistic fatigue damage estimation of embedded electronic solder joints under random vibration”, *Microelectronics Reliability*, vol. 78, pp. 249–257, 2017.

[KAM 20] KAMEL A., DAMMAK K., EL HAMI A. et al., “A modified hybrid method for a reliability-based design optimization applied to an offshore wind turbine”, *Mechanics of Advanced Materials and Structures*, pp. 1–14, 2020.

[KHA 01] KHARMANDA G., MOHAMED A., LEMAIRE M., “New hybrid formulation for reliability-based optimization of structures”, in *The Fourth World Congress of Structural and Multidisciplinary Optimization, WCSMO-4*, Dalian, 4–8 June 2001.

[KHA 02] KHARMANDA G., MOHAMED A., LEMAIRE M., “Efficient reliability-based design optimization using a hybrid space with application to finite element analysis”, *Structural and Multidisciplinary Optimization*, vol. 24, no. 3, pp. 233–245, 2002.

[KHA 03] KHARMANDA G., MOHSINE A., EL HAMI A., “Efficient reliability-based design optimization for dynamic structures”, in *The Fifth World Congress of Structural and Multidisciplinary Optimization, WCSMO-5*, Lido di Jesolo-Venice, 19–23 May 2003.

[KHA 04] KHARMANDA G., OLHOFF N., EL-HAMI A., “Optimum values of structural safety factors for a predefined reliability level with extension to multiple limit states”, *Structural and Multidisciplinary Optimization*, vol. 27, pp. 421–434, 2004.

[KHA 08] KHARMANDA G., MOHSINE A., MAKLOUFI A. et al., “Recent methodologies for reliability-based design optimization”, *International Journal for Simulation and Multidisciplinary Design Optimization*, vol. 2, no. 1, pp. 11–23, 2008.

[KHA 09] KHARMANDA G., SHARABATEY S., IBRAHIM H., et al., "Reliability-based design optimization using semi-numerical strategies for structural engineering applications", *International Journal of CAD/CAM*, vol. 9, no. 1, pp. 1–16, 2009.

[KHA 14] KHARMANDA G., IBRAHIM M.H., ABO AL-KHEER A. et al., "Reliability-based design optimization of shank chisel plough using optimum safety factor strategy", *Computers and Electronics in Agriculture*, vol. 109, pp. 162–171, 2014.

[KHA 16] KHARMANDA G., EL HAMI A., *Reliability in Biomechanics*. ISTE Ltd, London and John Wiley & Sons, New York, 2016.

[KHA 17] KHARMANDA G., EL HAMI A., *Biomechanics: Optimization, Uncertainties and Reliability*. ISTE Ltd, London and John Wiley & Sons, New York, 2017.

[KIM 00] KIM I.Y., KWAK B.M., "Design space optimization using a numerical design continuation method", *International Journal for Numerical Methods in Engineering*, vol. 53, pp. 1979–2002, 2000.

[KLE 92] KLEIBER M., HIEN T.D., *The Stochastique Finite Element Method*, John Wiley & Sons, Hoboken, NJ, 1992.

[KOU 20] KOULOU A., ZEMZAMI M., EL HAMI N. et al., "Optimization in collaborative information systems for an enhanced interoperability network", *International Journal for Simulation and Multidisciplinary Design Optimization*, vol. 11, p. 2019021, 2020.

[KOY 94] KOYLUOGLU H.U., NIELSEN S.R.K., "New approximations for SORM integrals", *Structurals, Structural Safety*, vol. 13, no. 4, pp. 235–246. 1994.

[KUS 97] KUSCHEL N., RACKWITZ "To basic problems in reliability-based structural optimization", *Mathematical Methods of Operation Research*, vol. 46, pp. 309–333, 1997.

[KUS 17] KUSANO I., BALDOMIR A., JURADO J.A. et al., "Reliability based design optimization of long-span bridges considering flutter", *Journal of Wind Engineering and Industrial Aerodynamics*, vol. 135, pp. 149–162, 2017.

[LAM 10] LAMBERT S., PAGNACCO E., KHALIJ L., "A probabilistic model for the fatigue reliability of structures under random loadings with phase shift effects", *International Journal of Fatigue*, vol. 32, no. 2, pp. 463–474, 2010.

[LED 17a] LEDOUX M., EL HAMI A., *Fluid Mechanics Analytical Methods*, ISTE Ltd, London and John Wiley & Sons, New York, 2017.

[LED 17b] LEDOUX M., EL HAMI A., *Compressible Flow, Propulsion and Digital Approaches in Fluid Dynamics*, ISTE Ltd, London and John Wiley & Sons, New York, 2017.

[LED 21a] LEDOUX M., EL HAMI A., *Heat Transfer 1: Conduction*, ISTE Ltd, London and John Wiley & Sons, New York, 2021.

[LED 21b] LEDOUX M., EL HAMI A., *Heat Transfer 2: Radiative Transfer*, ISTE Ltd, London and John Wiley & Sons, New York, 2021.

[LEE 02] LEE J.O., YANG Y.-S., Ruy W.-S., "A comparative study on reliability-index and target-performance-based probabilistic structural design optimization", *Computers and Structural*, vol. 80, pp. 257–267, 2002.

[LEM 05] LEMAIRE M., CHATEAUNEUF A., MITTEAU J., *Fiabilité des structures : couplage mécano-fiabiliste statique*, Hermès Sciences, Paris, 2005.

[LOP 11] LOPEZ R., HOLDORF D., LEMOSSE E.S., "An approach for the reliability based design optimization of laminated composite plates", *Engineering Optimization*, vol. 43, no. 10, pp. 1079–1094, 2011.

[LOU 07] LOUKUS A., SUBHASH G., IMANINEJAD, M., "An optimization of material properties and process parameters for tube hydroforming of aluminum extrusions", *Journal of Engineering Materials and Technology*, vol. 129, pp. 233–241, 2007.

[MAB 17] MABROUK I.B., EL HAMI A., WALHA L. et al., "Dynamic vibrations in wind energy systems: Application to vertical axis wind turbine", *Mechanical Systems and Signal Processing*, vol. 85, pp. 396–414, 2017.

[MAC 03] MACKERLE J., "Topology and shape optimization of structures using FEM and BEM: A bibliography (1999-2001)", *Finite Element in Analysis and Design*, vol. 39, pp. 243–253, 2003.

[MAH 99] MAHADEVAN S., SHI P., SALTER J., "Advanced reliability methods, AIAA/ASME/ASME/ASC structures", *Structural Dynamics and Materials Conference*, vol. 4, pp. 3026–3034, 1999.

[MAK 16] MAKHLOUFI A., AOUES Y., EL HAMI A., "Reliability based design optimization of wire bonding in power microelectronic devices", *Microsystem Technologies*, vol. 22, no. 12, pp. 2737–2748, 2016.

[MAN 13] MANSOURI M., Etude mécano-fiabiliste et réduction du modèle des problèmes vibro-acoustiques à paramètres aléatoires, Thesis, INSA Rouen Normandie, defended 22 April 2013.

[MIN 86] MINOUX M., *Mathematical Programming: Theory and Algorithms*, John Wiley & Sons, 1986.

[MOH 03] MOHSINE A., KHARMANDA G., EL HAMI A., "Integration of equivalent model within reliability-based design optimization for reducing freedom degrees in dynamic structures", in *Proceedings of 11th IFIP Conferences on Reliability and Optimization Systems*, Banff, 2–5 November, pp. 157–164, 2003.

[MOH 04] MOHSINE A., EL HAMI A., "Optimisation fiabiliste et méthodes de synthèses modales en dynamique des structures", in *14ème Congrès national de maîtrise des risques et sûreté de Fonctionnement Lambda-Mu 14*, Bourges, 12–14 October, pp. 668–690, 2004.

[MOH 05a] MOHSINE A., KHARMANDA G., EL HAMI A., "Improved hybrid method for reliability-based design optimization", in *12th WG 7.5 Working Conference on Reliability and Optimization of Structural Systems*, Aalborg, 22–25 May 2005.

[MOH 05b] MOHSINE A., KHARMANDA G., EL HAMI A., "Reliability-based design optimisation study using normal and lognormal distributions with applications to dynamic structures", in *9th International Conference on Structural Safety and Reliability*, ICOSSAR05, Rome, 19–23 June 2005.

[MOH 10] MOHSINE A., EL HAMI A., "A robust study of reliability-based optimization methods under eigen-frequency", *Computer Methods in Applied Mechanics and Engineering*, vol. 199, nos 17–20, pp. 1006–1018, 2010.

[MOR 02] MORO T., EL HAMI A., EL MOUDNI A., "Reliability analysis of a mechanical contact between deformable solids", *Probabilistic Engineering Mechanics*, vol. 17, no. 3, pp. 227–232, 2002.

[NAK 01] NAKANISHI Y., "Application of homology theory to topology optimization of three-dimensional structures using genetic algorithms", *Computer Methods in Applied Mechanics and Engineering*, vol. 190, pp. 3849–3863, 2001.

[PAP 02] PAPADRAKADIS M., LAGAROS N., "Reliability-based structural optimization using neutral networks and Monte Carlo simulation", *Computer Methods in Applied Mechanics and Engineering*, vol. 191, pp. 3491–35007, 2002.

[PIT 01] PITOISET X., RYCHLIK I., PREUMONT A., "Spectral methods to estimate local multiaxial fatigue failure for structures undergoing random vibrations", *Fatigue & Fracture of Engineering Materials & Structures*, vol. 24, no. 11, pp. 715–727, 2001.

[PRA 75] PRASTHOFER P.H., BEADLE C.H., "Dynamic response of structures with statistical uncertainties in their stiffness", *Journal of Sound and Vibration*, vol. 42, no. 4, pp. 477–493, 1975.

[PRO 01] PROOS K., STEVEN G., SWAIN M. et al., "Preliminary studies on the optimum shape of dental bridges". *Computer Methods in Biomechanics and Biomedical Engineering*, vol. 4, no. 1, pp. 77–92, 2001.

[RAC 78] RACKWITZ R., FIESSLER B., "Structural reliability under combined random load sequence", *Computers and Structures*, vol. 9, no. 5, pp. 489–494, 1978.

[RAD 07] RADI B., El HAMI A., "Reliability analysis of the metal forming process", *Mathematical and Computer Modelling*, vol. 45, nos 3–4, pp. 431–439, 2007.

[RAD 10] RADI B. et al., "Reliability-based design optimization of an ultrasonic motor by new approach", *Simulation and Multidisciplinary Design Optimization*, vol. 4, pp. 71–77, 2010.

[RAD 16] RADI B., EL HAMI A., *Material Forming Processes: Simulation, Drawing, Hydroforming and Additive Manufacturing*, ISTE Ltd, London and John Wiley & Sons, New York, 2016.

[RAI 02] RAIS-ROHANI M., SINGH M.N., "Efficient response surface approach for reliability estimation of composite structures", in *Proceedings of the 9th AIAA/ ISSMO Symposium on Multidisciplinary Analysis and Optimization*, Atlanta, GA, 4–6 September 2002.

[REY 99] REYNOLDS D., MCCONNACHIE J., BETTESS P. et al., "Reverse adaptivity – A new evolutionary tool for structural optimisation", *International Journal for Numerical Method in Engineering*, vol. 45, no. 5, pp. 529–552, 1999.

[RHO 01] RHOUAS S., BOUCHEKOURTE M., EL HAMI N., "Optimization of the impact measurement of market structure on liquidity and volatility", *International Journal for Simulation and Multidisciplinary Design Optimization*, vol. 2022, no. 13, p. 9, 2001.

[ROY 01] ROYSET J.O., DER KIUREGHIAN A., POLAK E., "Reliability-based optimal design of series structural systems", *Journal of Engineering Mechanics*, vol. 127, pp. 607–614, 2001.

[ROZ 93] ROZVANY G.I.N., *Optimisation of Large Structural Systems*, Kluwer, Dordrecht, 1993.

[SAH 13] SAHA S., SEPAHVAND K., MATSAGAR V. et al., "Stochastic analysis of base-isolated liquid storage tanks with uncertain isolator parameters under random excitation", *Engineering Structures*, vol. 57, pp. 465–474, 2013.

[SAR 11] SARSRI D., AZRAR L., JEBBOURI A. et al., "Component mode synthesis and polynomial chaos expansions for stochastic frequency functions of large linear FE models", *Computers and Structures*, vol. 89, nos 3–4, pp. 346–356, 2011.

[SCH 60] SCHMIT L.A., "Structural design by systematic synthesis", in *Proceedings of the Second ASCE Conference on Electronic Computation*, Pittsburgh, PA, pp. 105–122, 1960.

[SCH 84] SCHMIT L.A., "Structural optimization-some key ideas and insights", in ATREK E., GALLAGHER R.H., RAGSDELL K.M. et al. (eds), *New Directions in Optimum Structural Design*, John Wiley & Sons, Chicester, pp. 1–45, 1984.

[SCH 01] SCHUËLLER G.I., "Comutational stochasic mechanics – recent advances", *Computers and Structures*, vol. 79, pp. 2225–2234, 2001.

[SEP 15] SEPAHVAND K., SCHEFFLER M., MARBURG S., "Uncertainty quantification in natural frequencies and radiated acoustic power of composite plates: Analytical and experimental investigation", *Applied Acoustics*, vol. 87, pp. 23–29, 2015.

[SHA 97] SHAO S., MUROSTSU Y., "Structural reliability analysis using a neutral network", *JAME, International Journal, Series A*, vol. 40, pp. 242–246, 1997.

[SHI 72] SHINOZUKA M., "Monte Carlo solution of structural dynamics", *International Journal Computer Structural*, vol. 2, pp. 855–874, 1972.

[SHI 83] SHINOUZUKA M., "Basic analysis of structural safety", *Journal of the Structural Division*, ASCE, vol. 109, no. 3, pp. 721–740, 1983.

[SHU 07] SHU-HUI L., YANG B., ZHANG W.G. et al., "Loading path prediction for tube hydroforming process using a fuzzy control strategy", *Materials and Design*, vol. 29, no. 6, pp. 1110–1116, 2007.

[SOI 93] SOIZE C., *Méthodes mathématiques en analyse du signal*, Masson, Paris, 1993.

[SPA 08] SPANOS P., KONTSOS A., "A multiscale monte carlo finite element method for determining mechanical properties of polymer nanocomposites", *Probabilistic Engineering Mechanics*, vol. 23, no. 4, pp. 456–470, 2008.

[STR 01] STRANO M., JIRATHEARANAT S., ALTAN T., "Adaptive FEM simulation for tube hydroforming: A geometry-based approach for wrinkle detection", *Manufacturing Technology*, vol. 50, pp. 185–190, 2001.

[SU 02] SU B., RAIS-ROHANI M., SINGH M.N., "Reliability-based optimization of anisotropic cylindrical shells with response surface approximations of buckling instability", in *Proceedings of the 43rd AiAA/ASME/ASCE/AHS/ASC Structural Dynamics, and Material Conference*, Denver, CO, AIAA-2002-1386, 22–25 April 2002.

[TRO 87] TROMPETTE P., FLEURY C., KNOPF-LENOIR C., Optimisation des structures : approche de l'ingénieur, Course Note, Paris 1987.

[TVE 90] TVEDT L., "Distribution of quadratic forms in normal space: Application to structural reliability", *Journal of Engineering Mechanics*, vol. 116, pp. 1183–1197, 1990.

[WAN 96] WANG L., HANDHI R.V., "System reliability analysis optimization with the use of higher-order approximations", *AIAA-96-4043-CP*, pp. 579–591, 1996.

[WEB 99] WEBER B., Fatigue multiaxiale des structures industrielles sous chargement quelconque, Thesis, INSA Lyon, 1999.

[WIE 38] WIENER N., "The homogeneous chaos", *American Journal of Mathematics*, vol. 60, no. 4, pp. 897–936, 1938.

[XIA 15] XIA B., YU D., "Optimization based on reliability and confidence interval design for the structural-acoustic system with interval probabilistic variables", *Journal of Sound and Vibration*, vol. 336, pp. 1–15, 2015.

[YAI 17] YAICH A., EL HAMI A., WALHA L. et al., "Local multiaxial fatigue damage estimation for structures under random vibrations", *Finite Elements in Analysis and Design*, vol. 132, pp. 1–7, 2017.

[YAI 18] YAICH A., KHARMANDA G., EL HAMI A. et al., "Reliability based design optimization for multiaxial fatigue damage analysis using robust hybrid method", *Journal of Mechanics*, vol. 34, no. 5, pp. 551–566, 2018.

[YAN 91] YANG J.S., NIKOLAIDIS E., "Design of aircraft wings subject to gust loads: A safety index based approach", *AIAA Journal*, vol. 29, pp. 804–812, 1991.

[ZAD 65] ZADEH L.A., "Fuzzy set", *Information and Control*, vol. 8, pp. 338–353, 1965.

[ZAD 95] ZADEH L.A., "Soft computing and fuzzy Logic", *IEEE Software*, November, vol. 11, no. 6, November, pp. 48–56, 1994.

[ZEM 19a] ZEMZAMI M., ELHAMI N., ITMI M. et al., "A modified particle swarm optimization algorithm linking dynamic neighborhood topology to parallel computation", *International Journal of Advanced Trends in Computer Science and Engineering*, vol. 8, no. 2, pp. 112–118, 2019.

[ZEM 19b] ZEMZAMI M., KOULOU A., ELHAMI N. et al., "Interoperability optimization using a modified PSO algorithm", *International Journal of Advanced Trends in Computer Science and Engineering*, vol. 8, no. 2, pp. 101–107, 2019.

[ZEM 20] ZEMZAMI M., EL HAMI N., ITMI M. et al., "A comparative study of three new parallel models based on the PSO algorithm", *International Journal for Simulation and Multidisciplinary Design Optimization*, vol. 11, pp. 2019022, 2020.

[ZHA 92] ZHANG W.H., Calcul des sensibilités et optimisation de forme par la méthode des élément finis, Thesis, Université de Liège, 1992.

[ZHA 98a] ZHAO Y.G., ONO T., "New approximations for SORM: Part 1 and 2", *Journal of Engineering Mechanics*, vol. 125, pp. 79–93, 1998.

[ZHA 98b] ZHAO Y.G., ONO, T., "System reliability evaluation of ductile frame structure", *Journal of Structural Engineering*, vol. 124, pp. 678–685, 1998.

Index

R, S

U, V, W

Other titles from

iSTE

in

Mechanical Engineering and Solid Mechanics

2025

LEROY Maurice

Rheology, Physical and Mechanical Behavior of Materials 4: Rigidity and Resistance of Composite Materials, Sizings of Laminate

Rheology, Physical and Mechanical Behavior of Materials 5: Fractures and Fatigue of Metal and Composites

EL HAMI Abdelkhalak

Methods and Applications of Artificial Intelligence: Dynamic Response, Learning, Random Forest, Linear Regression, Interoperability, Additive Manufacturing and Mechatronics
(Artificial Intelligence in Mechanics Set – Volume 2)

2024

LE VAN Anh

Contact in Structural Mechanics: A Weighted Residual Approach

LEROY Maurice

Rheology, Physical and Mechanical Behavior of Materials 2: Working at High Strain Rates, Forming, Sintering, Joining and Welding

Rheology, Physical and Mechanical Behavior of Materials 3: Rigidity and Resistance of Materials, Sizings, Pieces and Structures

SAOUAB Abdelghani, CHAMPMARTIN Stéphane, KHALID NACIRI Jaafar

Some Complex Phenomena in Fluid and Solid Mechanics

(Mathematical and Mechanical Engineering Set – Volume 13)

2023

EL HAMI Abdelkhalak

Uncertainty and Artificial Intelligence: Additive Manufacturing, Vibratory Control, Agro-composite, Mechatronics

(Artificial Intelligence in Mechanics Set – Volume 1)

EL HAMI Abdelkhalak, EID Mohamed

Reliability-based Modeling of System Performance

(Reliability of Multiphysical Systems Set – Volume 19)

EL HAMI Abdelkhalak, DELAUX David, GRZESKOWIAK Henri

Applied Reliability for Industry 1: Predictive Reliability for the Automobile, Aeronautics, Defense, Medical, Marine and Space Industries

(Reliability of Multiphysical Systems Set – Volume 16)

Applied Reliability for Industry 2: Experimental Reliability for the Automobile, Aeronautics, Defense, Medical, Marine and Space Industries

(Reliability of Multiphysical Systems Set – Volume 17)

Applied Reliability for Industry 3: Operational Reliability for the Automobile, Aeronautics, Defense, Medical, Marine and Space Industries

(Reliability of Multiphysical Systems Set – Volume 18)

LEROY Maurice

Rheology, Physical and Mechanical Behavior of Materials 1: Physical Mechanisms of Deformation and Dynamic Behavior

2022

BAYLE Franck
Product Maturity 1: Theoretical Principles and Industrial Applications
(Reliability of Multiphysical Systems Set – Volume 12)
Product Maturity 2: Principles and Illustrations
(Reliability of Multiphysical Systems Set – Volume 13)

EL HAMI Abdelkhalak, DELAUX David, GRZESKOWIAK Henri
Reliability and Physics-of-Healthy in Mechatronics
(Reliability of Multiphysical Systems Set – Volume 15)

LANNOY André
Reliability of Nuclear Power Plants: Methods, Data and Applications
(Reliability of Multiphysical Systems Set – Volume 14)

LEDOUX Michel, EL HAMI Abdelkhalak
Heat Transfer 3: Convection, Fundamentals and Monophasic Flows
(Mathematical and Mechanical Engineering Set – Volume 11)
Heat Transfer 4: Convection, Two-Phase Flows and Special Problems
(Mathematical and Mechanical Engineering Set – Volume 12)

PLANCHETTE Guy
Cindynics, The Science of Danger: A Wake-up Call
(Reliability of Multiphysical Systems Set – Volume 11)

2021

CHALLAMEL Noël, KAPLUNOV Julius, TAKEWAKI Izuru
Modern Trends in Structural and Solid Mechanics 1: Static and Stability
Modern Trends in Structural and Solid Mechanics 2: Vibrations
Modern Trends in Structural and Solid Mechanics 3: Non-deterministic Mechanics

DAHOO Pierre Richard, POUGNET Philippe, EL HAMI Abdelkhalak
Applications and Metrology at Nanometer Scale 1: Smart Materials, Electromagnetic Waves and Uncertainties
(Reliability of Multiphysical Systems Set – Volume 9)
Applications and Metrology at Nanometer Scale 2: Measurement Systems, Quantum Engineering and RBDO Method
(Reliability of Multiphysical Systems Set – Volume 10)

LEDOUX Michel, EL HAMI Abdelkhalak
Heat Transfer 1: Conduction
(Mathematical and Mechanical Engineering Set – Volume 9)
Heat Transfer 2: Radiative Transfer
(Mathematical and Mechanical Engineering Set – Volume 10)

2020

SALENÇON Jean
Elastoplastic Modeling

2019

BAYLE Franck
Reliability of Maintained Systems Subjected to Wear Failure Mechanisms: Theory and Applications
(Reliability of Multiphysical Systems Set – Volume 8)

BEN KAHLA Rabeb, BARKAOUI Abdelwahed, MERZOUKI Tarek
Finite Element Method and Medical Imaging Techniques in Bone Biomechanics
(Mathematical and Mechanical Engineering Set – Volume 8)

IONESCU Ioan R., QUEYREAU Sylvain, PICU Catalin R., SALMAN Oguz Umut
Mechanics and Physics of Solids at Micro- and Nano-Scales

LE VAN Anh, BOUZIDI Rabah
Lagrangian Mechanics: An Advanced Analytical Approach

MICHELITSCH Thomas, PÉREZ RIASCOS Alejandro, COLLET Bernard, NOWAKOWSKI Andrzej, NICOLLEAU Franck
Fractional Dynamics on Networks and Lattices

SALENÇON Jean
Viscoelastic Modeling for Structural Analysis

VÉNIZÉLOS Georges, EL HAMI Abdelkhalak
Movement Equations 5: Dynamics of a Set of Solids
(Non-deformable Solid Mechanics Set – Volume 5)

2018

BOREL Michel, VÉNIZÉLOS Georges
Movement Equations 4: Equilibriums and Small Movements
(Non-deformable Solid Mechanics Set – Volume 4)

FROSSARD Etienne
Granular Geomaterials Dissipative Mechanics: Theory and Applications in Civil Engineering

RADI Bouchaib, EL HAMI Abdelkhalak
Advanced Numerical Methods with Matlab® 1: Function Approximation and System Resolution
(Mathematical and Mechanical Engineering Set – Volume 6)
Advanced Numerical Methods with Matlab® 2: Resolution of Nonlinear, Differential and Partial Differential Equations
(Mathematical and Mechanical Engineering Set – Volume 7)

SALENÇON Jean
Virtual Work Approach to Mechanical Modeling

2017

BOREL Michel, VÉNIZÉLOS Georges
Movement Equations 2: Mathematical and Methodological Supplements
(Non-deformable Solid Mechanics Set – Volume 2)
Movement Equations 3: Dynamics and Fundamental Principle
(Non-deformable Solid Mechanics Set – Volume 3)

BOUVET Christophe
Mechanics of Aeronautical Solids, Materials and Structures
Mechanics of Aeronautical Composite Materials

BRANCHERIE Delphine, FEISSEL Pierre, BOUVIER Salima, IBRAHIMBEGOVIC Adnan
From Microstructure Investigations to Multiscale Modeling: Bridging the Gap

CHEBEL-MORELLO Brigitte, NICOD Jean-Marc, VARNIER Christophe
From Prognostics and Health Systems Management to Predictive Maintenance 2: Knowledge, Traceability and Decision
(Reliability of Multiphysical Systems Set – Volume 7)

EL HAMI Abdelkhalak, RADI Bouchaib
Dynamics of Large Structures and Inverse Problems
(Mathematical and Mechanical Engineering Set – Volume 5)
Fluid-Structure Interactions and Uncertainties: Ansys and Fluent Tools
(Reliability of Multiphysical Systems Set – Volume 6)

KHARMANDA Ghias, EL HAMI Abdelkhalak
Biomechanics: Optimization, Uncertainties and Reliability
(Reliability of Multiphysical Systems Set – Volume 5)

LEDOUX Michel, EL HAMI Abdelkhalak
Compressible Flow Propulsion and Digital Approaches in Fluid Mechanics
(Mathematical and Mechanical Engineering Set – Volume 4)
Fluid Mechanics: Analytical Methods
(Mathematical and Mechanical Engineering Set – Volume 3)

MORI Yvon
Mechanical Vibrations: Applications to Equipment

2016

BOREL Michel, VÉNIZÉLOS Georges
Movement Equations 1: Location, Kinematics and Kinetics
(Non-deformable Solid Mechanics Set – Volume 1)

BOYARD Nicolas
Heat Transfer in Polymer Composite Materials

CARDON Alain, ITMI Mhamed
New Autonomous Systems
(Reliability of Multiphysical Systems Set – Volume 1)

DAHOO Pierre Richard, POUGNET Philippe, EL HAMI Abdelkhalak
Nanometer-scale Defect Detection Using Polarized Light
(Reliability of Multiphysical Systems Set – Volume 2)

DORMIEUX Luc, KONDO Djimédo
Micromechanics of Fracture and Damage
(Micromechanics Set – Volume 1)

EL HAMI Abdelkhalak, RADI Bouchaib
Stochastic Dynamics of Structures
(Mathematical and Mechanical Engineering Set – Volume 2)

GOURIVEAU Rafael, MEDJAHER Kamal, ZERHOUNI Noureddine
From Prognostics and Health Systems Management to Predictive Maintenance 1: Monitoring and Prognostics
(Reliability of Multiphysical Systems Set – Volume 4)

KHARMANDA Ghias, EL HAMI Abdelkhalak
Reliability in Biomechanics
(Reliability of Multiphysical Systems Set –Volume 3)

MOLIMARD Jérôme
Experimental Mechanics of Solids and Structures

RADI Bouchaib, EL HAMI Abdelkhalak
Material Forming Processes: Simulation, Drawing, Hydroforming and Additive Manufacturing
(Mathematical and Mechanical Engineering Set – Volume 1)

DE SAXCÉ Géry, VALLÉE Claude
Galilean Mechanics and Thermodynamics of Continua

2015

JEBAHI Mohamed, ANDRE Damien, TERREROS Inigo, IORDANOFF Ivan
Discrete Element Method to Model 3D Continuous Materials
Discrete Element Model and Simulation of Continuous Materials Behavior Set – Volume 1

JEBAHI Mohamed, DAU Frédéric, CHARLES Jean-Luc, IORDANOFF Ivan
Discrete-continuum Coupling Method to Simulate Highly Dynamic Multi-scale Problems: Simulation of Laser-induced Damage in Silica Glass
Discrete Element Model and Simulation of Continuous Materials Behavior Set – Volume 2

KARLIČIĆ Danilo, MURMU Tony, ADHIKARI Sondipon, MCCARTHY Michael
Non-local Structural Mechanics

SAB Karam, LEBÉE Arthur
Homogenization of Heterogeneous Thin and Thick Plates

2014

ATANACKOVIC M. Teodor, PILIPOVIC Stevan, STANKOVIC Bogoljub, ZORICA Dusan
Fractional Calculus with Applications in Mechanics: Vibrations and Diffusion Processes
Fractional Calculus with Applications in Mechanics: Wave Propagation, Impact and Variational Principles

CIBLAC Thierry, MOREL Jean-Claude
Sustainable Masonry: Stability and Behavior of Structures

ILANKO Sinniah, MONTERRUBIO Luis E., MOCHIDA Yusuke
The Rayleigh–Ritz Method for Structural Analysis

LALANNE Christian
Mechanical Vibration and Shock Analysis – 5-volume series – 3rd edition
Sinusoidal Vibration – Volume 1
Mechanical Shock – Volume 2
Random Vibration – Volume 3
Fatigue Damage – Volume 4
Specification Development – Volume 5

LEMAIRE Maurice
Uncertainty and Mechanics

2013

ADHIKARI Sondipon
Structural Dynamic Analysis with Generalized Damping Models: Analysis

ADHIKARI Sondipon
Structural Dynamic Analysis with Generalized Damping Models: Identification

BAILLY Patrice
Materials and Structures under Shock and Impact

BASTIEN Jérôme, BERNARDIN Frédéric, LAMARQUE Claude-Henri
Non-smooth Deterministic or Stochastic Discrete Dynamical Systems: Applications to Models with Friction or Impact

EL HAMI Abdelkhalak, RADI Bouchaib
Uncertainty and Optimization in Structural Mechanics

KIRILLOV Oleg N., PELINOVSKY Dmitry E.
Nonlinear Physical Systems: Spectral Analysis, Stability and Bifurcations

LUONGO Angelo, ZULLI Daniele
Mathematical Models of Beams and Cables

SALENÇON Jean
Yield Design

2012

DAVIM J. Paulo
Mechanical Engineering Education

DUPEUX Michel, BRACCINI Muriel
Mechanics of Solid Interfaces

ELISHAKOFF Isaac *et al.*
Carbon Nanotubes and Nanosensors: Vibration, Buckling and Ballistic Impact

GRÉDIAC Michel, HILD François
Full-Field Measurements and Identification in Solid Mechanics

GROUS Ammar
Fracture Mechanics – 3-volume series
Analysis of Reliability and Quality Control – Volume 1
Applied Reliability – Volume 2
Applied Quality Control – Volume 3

RECHO Naman
Fracture Mechanics and Crack Growth

2011

KRYSINSKI Tomasz, MALBURET François
Mechanical Instability

LAMBERT Stéphane, NICOT François
Rockfall Engineering

SOUSTELLE Michel
An Introduction to Chemical Kinetics

2010

BREITKOPF Piotr, FILOMENO COELHO Rajan
Multidisciplinary Design Optimization in Computational Mechanics

DAVIM J. Paulo
Biotribology

PAULTRE Patrick
Dynamics of Structures

SOUSTELLE Michel
Handbook of Heterogenous Kinetics

2009

BERLIOZ Alain, TROMPETTE Philippe
Solid Mechanics using the Finite Element Method

LEMAIRE Maurice
Structural Reliability

2007

GIRARD Alain, ROY Nicolas
Structural Dynamics in Industry

GUINEBRETIÈRE René
X-ray Diffraction by Polycrystalline Materials

KRYSINSKI Tomasz, MALBURET François
Mechanical Vibrations

KUNDU Tribikram
Advanced Ultrasonic Methods for Material and Structure Inspection

SIH George C., NAÏT-ABDELAZIZ Moussa, VU-KHANH Toan
Particle and Continuum Aspects of Mesomechanics

Printed and bound by CPI Group (UK) Ltd, Croydon, CR0 4YY

24/08/2026

14947835-0003